How to Protect What's Yours

How to Protect What's Yours

Nancy Golonka
Edited by Rush Loving, Jr.

ACROPOLIS BOOKS LTD.
Washington, D.C.

Library of Congress Cataloging in Publication Data

Golonka, Nancy, 1946-
  How to protect what's yours.

  1. Insurance—Handbooks, manuals, etc. I. Loving, Rush. II. Title.
  HG8061.G64 1983   368'.002'02   83-3825
  ISBN 0-87491-557-0
  ISBN 0-87491-560-0 (pbk.)

© Copyright 1983,
Acropolis Books Ltd. and the Insurance Information Institute

All rights reserved. Except for the inclusion of brief quotations in a review, no part of this book may be reproduced or utilized in any form or by any means, electronic or mechanical, including photocopying, recording, or by any information storage and retrieval system, without permission in writing from the publisher.

ACROPOLIS BOOKS LTD.
Colortone Building, 2400 17th Street, N.W.
Washington, D.C. 20009

Printed in the United States of America by
Colortone Press
Creative Graphics Inc.
Washington, D.C. 20009

ACKNOWLEDGMENTS

The support and the contributions of some very special people have helped to make this book possible.

Many agents, company officers and representatives of the property, casualty, life and health insurance industry graciously shared anecdotes from their experiences in working with consumers. Their help enabled me to include a number of real-life examples to illustrate how insurance works and how policyholders can make it work better while saving money at the same time. To protect individual privacy, the names of the people involved have been changed, except for the story of the devastating tornado in Wichita Falls, Texas.

I am grateful to Charles J. O'Connor, industry affairs officer at the Kemper Group, and Alfred I. Jaffe, associate professor at The College of Insurance, for their thoroughness and patience in reviewing the manuscript.

For their enormous professional talents, I also want to thank my colleagues at the Insurance Information Institute. Bernard Kaapcke devoted many weeks to detailed research and his unique skills were vital in organizing the contents of the book. John Craigie gave valuable guidance on the auto and homeowners insurance chapters. Samuel Schiff provided the benefit of his considerable expertise to the section on protecting the family nest egg. And Mary Zavada assisted most ably with the life and health chapters.

It was the vision of the Institute's senior officers which made possible the creation of a comprehensive consumer affairs program, of which this book now becomes a significant part. For their enthusiastic support and interest, I want to say a special word of gratitude to Mechlin D. Moore, president of the Institute, and Charles C. Clarke, executive vice president.

Daniel Poole, the Institute's vice president for publications, is to be commended for managing so well the giant task of coordinating the work that went into this project. And thanks is due to Lilia Piotrowski for her efforts and endurance in typing the manuscript.

Special appreciation is due, too, to Rush Loving, Jr., for his excellent editing and for adding the finishing touches which make the book so interesting for the reader.

Nancy Golonka
New York, New York

CONTENTS

Introduction 1

Part One
PROTECTING YOUR HOUSE AND POSSESSIONS

1 **When Everything Disappears in a Cloud** 6
   Tale of a tornado . . . Picking up the pieces . . . Financial impact of a disaster . . . Protecting your assets through insurance.

2 **Keeping a Roof Over Your Head** 13
   Understanding your homeowners insurance policy . . . When you have a business in your home . . . Insuring your personal belongings . . . What your homeowners insurance will pay for . . . What isn't covered . . . Insuring a mobile home.

3 **Getting Full Protection Without Overspending** 26
   How to avoid buying more insurance than you need . . . Guarding against inflation . . . Ways to save on home insurance . . . Discounts. . . Keeping an inventory of your property.

4 **How to Keep Your Home From Bankrupting You** 36
   Why you need liability insurance . . . Help if you're hurt at home . . . If you damage someone else's property . . . If you rent your property . . . Protection for your boat . . . Compensation for household employees.

5 **Watching Out for Your Personal Possessions** 46
   Burglary and theft coverage . . . Personal articles insurance.

6 **Disasters Mother Never Told You About** 52
   Protecting the title to your house . . . Flood and earthquake . . . Crime insurance . . . Insurance for "uninsurable" property . . . For beach and windstorm areas.

7 **Insurance for the Barnyard** 62
   Protection for farmers and ranchers . . . Farm equipment and livestock . . . Farm buildings . . . Farmers' liability . . . Farm vehicles . . . Crop-hail insurance.

8 What to Do When You Have a Claim 67
   First steps . . . Reporting your loss . . . Insurance adjusting . . . How your settlement is determined . . . Liability situations . . . Medical payments . . . If you disagree with the insurance company's offer.

9 How to Avoid Disaster 76
   How to protect against fire . . . What to do about windstorms . . . How to handle a hurricane . . . How to keep the burglar away . . . Home safety.

Part Two
PROTECTING WHAT'S YOURS ON THE ROAD

10 Your Auto Insurance 95
   Why you need it . . . Drivers' liability . . . Liability for personal injuries . . . For property damage . . . Medical payments . . . Insurance against uninsured motorists . . . Collision insurance . . . Fire, theft and other hazards . . . Other vehicles . . . Legal requirements . . . Questions that arise.

11 All About No-Fault 117
   The trouble with lawsuits . . . Keeping out of court with no-fault . . . How no-fault insurance works.

12 How to Hold Your Auto Premiums Down 125
   Effects of inflation on auto insurance . . . How your premium is figured . . . How to save on auto insurance . . . Discounts . . . Your driving record.

13 What If the Company Turns You Down? 145
   Anyone can get auto insurance . . . The "non-standard" insurance market . . . Assigned risk plans and other plans for the hard-to-insure . . . The way back to regular insurance.

14 How to Protect Yourself After an Accident 152
   When you have a crash . . . Steps to take . . . How to put your insurance to work. . . Aid for the hit-and-run victim.

Part Three
YOUR LIFE AND HEALTH

15 Protecting Your Estate 160
Safeguarding your dependents through life insurance . . .
Saving through life insurance . . . Term insurance . . .
Whole life insurance . . . Special policies . . . Group life . . .
Variable life . . . Universal life . . . Annuities . . . Dividends
from life insurance . . . Shopping for a policy.

16 When Sickness Strikes at Your Savings 176
Health insurance . . . Group plans . . . Health Maintenance
Organizations . . . Health service contractors . . . Your
coverages under health insurance . . . Major medical plans
. . . Dental insurance . . . What to look for in a health policy
. . . Disability income . . . Medicare and Medicaid . . . How
to keep down your medical costs.

Part Four
OTHER THINGS YOU SHOULD KNOW

17 Guarding Against Everything Else 194
You can insure almost anything . . . Where to get insurance
on unusual risks . . . The "excess and surplus" insurance
market . . . Lloyd's of London.

18 When You're the Victim 202
How other people's liability insurance protects you . . . How
to file a claim when you're the victim . . . Workers'
compensation . . . Other help for the victim.

Part Five
PROTECTING YOUR NEST EGG

19 When Recession Comes 212
The threat of unemployment . . . Guarding your job skills
. . . Handling debt . . . Financial priorities . . . Dealing with
inflation.

20 Keeping Your Nest Egg From Breaking 216
Helping your money grow . . . Savings accounts . . . Money market accounts . . . Common stocks . . . Managing an investment portfolio . . . Bonds . . . Precious metals and commodities . . . Art works and collectibles . . . Real estate.

21 Letting the Experts Do It 239
Advantages of a mutual fund . . . Money market funds . . . How to choose a fund . . . The magic of dollar cost averaging . . . Accumulation plans . . . Caring for your fund.

22 Retiring Without Pensioning Off What's Yours 254
Planning for retirement . . . Social Security . . . Company pensions . . . Employee savings plans . . . Individual Retirement Accounts and Keogh Plans . . . Annuities.

23 Parting Advice 266
How to shop for insurance . . . Planning your program . . . Getting professional advice . . . How to become an informed consumer . . . Ways to save on insurance . . . Keeping up with change in your life . . . Eliminating insurance you don't need . . . Keeping records . . . Knowing your resources.

APPENDICES

A How to Read Your Insurance Policy 284
B Your Choice of Homeowners Policies 292

INTRODUCTION

DON'T MISSPEND YOUR PROTECTION DOLLAR

Eighty percent of all Americans who buy insurance misspend some of their premium dollars by buying the wrong kinds of coverage. Often they buy insurance they don't need, and fail to buy insurance they do need.

Many families spend as much as $100 a year for double coverages against the same disaster, yet they fail to spend an extra $85 or so for $1 million of crucial "umbrella" liability insurance to back up their auto and homeowners policies and save them from being wiped out in a lawsuit. Recent surveys showed that in one California forest fire in 1977, 96 percent of the owners who lost their homes were underinsured, and in a 1982 fire about 40 percent had no coverage at all.

Fires, floods, lawsuits, thefts, inflation, the tax collector and a whole apocalypse of other perils threaten and even destroy people's possessions. Whether victims of physical disasters or economic erosion, many people are ill-prepared and poorly protected. And their losses frequently are shattering.

For instance, if recession causes you to lose your job, you could lose your home as well. Or if inflation runs at annual rates of 10 percent for as brief a time as five years, it can cut the real value of your retirement nest egg more than half. Or if someone falls on your sidewalk and breaks a hip irreparably, the resulting lawsuit could snatch away almost all of your possessions—and perhaps much of your future income as well. So it's crucial to protect what's yours.

That's why the Insurance Information Institute, an educational organization representing many of the nation's property and casualty insurance companies, sponsored the research that led to the publication of this book. Because it's imperative that you be informed about your insurance, the Institute also has established a toll-free consumer hotline to help you get answers to questions you may have about your insurance—especially your automobile and homeowners insurance. Just call 800-221-4954 (in New York State, call collect 212-669-9200).

To get your money's worth, you need to know how insurance works and what kinds you require for your particular possessions and lifestyle. That done, you should sit down with your insurance agent or company representative and look at all those needs and tailor your coverages to fit them.

But that's not all! Just as you must stay abreast of changes in the stock market when you invest, you also should keep your insurance agent informed of any changes in your life that can affect your insurance. But you can't do that intelligently or profitably unless you understand what different types of insurance do and what your particular coverages provide you.

Surveys show, for instance, that the most frequent complaint heard from consumers is that auto and homeowners insurance costs too much. One reason is that people often don't know enough about what they're buying. They are unable, therefore, to explain sufficiently to their agents precisely what their insurance needs are so the agent can recommend and tailor for them the right combination of coverages.

A lot of people pay extra to have only a $200 deductible on their auto coverage or homeowners insurance. For a considerable saving, they could take a $500 deductible if they're capable of covering the extra risk out of their own bank accounts. And they often might use the saving or part of it to raise the upper limits of their coverage. Say you have medical insurance with a $200 deductible and you raise it to

$1,000. You might well find that you have only a $25,000 maximum coverage in your policy. For just about the same amount you're saving by raising the deductible, you'll be able to raise the total coverage on your doctor's bills.

Of course, this book isn't just about insurance. Fire and theft are only two of the perils that threaten what's yours. Inflation can eat away at your nest egg, and the tax collector can snatch away some of your retirement income. We haven't stopped at insurance but have made this book a guide to protecting all of your possessions, whether they be homes, cars, furniture or cash. We don't claim to offer an all-encompassing guide to investments or retirement. But we do provide you with some good common-sense advice on how to protect your nest egg and your retirement income. And in the process, we offer you plenty of practical tips on how to protect the money in your pocket by getting the best protection at the lowest cost.

In short, this book is designed to help you save your money and make better use of it while you protect your worldly goods. If you read this book and keep it as a reference you not only will be far better equipped to buy proper insurance coverage and save premium dollars, but you will be spared costly losses later from disasters you thought were covered but really weren't. That's why this book was written. And that's why you should read it.

PART ONE

Protecting Your House And Possessions

CHAPTER 1

WHEN EVERYTHING DISAPPEARS IN A CLOUD

Great, muscular clouds of battleship gray, their bottoms tinged with black, boiled overhead. They rolled across the plains and over the town ominously, but they were a common springtime sight in north Texas, and the people were used to them. Nevertheless, by nightfall, Wichita Falls, a town of 97,000 souls that lies about 30 miles south of the Oklahoma line, would be hit by a disaster that would leave 45 dead and hundreds injured. And 20,000 people would lose their homes. Men, women and children would be unable to sleep that evening and for many nights to come because of trauma and emotional aftershocks.

At 4:30 p.m., a weather watcher 50 miles away reported a small funnel heading northeast toward the city. The National Weather Service upgraded the tornado watch to a warning a few minutes after 5 p.m. The residents of Wichita Falls were going about their business despite the blustery winds and humid air. Some were doing their Easter shopping, others driving home. The news on their car radios was dominated by troubles in Nicaragua and Iran and at Three Mile Island.

Louise Covington, a vivacious secretary, her hair just beginning to gray, arrived at her small frame home in the suburb of Faith Village and started fixing dinner. While she worked, Mrs. Covington was half-listening to the police band on the kitchen radio. Shortly before 6 p.m. a dispatcher's voice suddenly broke the monotone of routine messages. "We got two, maybe three, funnels over the CertainTeed," the dispatcher declared. Mrs. Covington dropped her mixing spoon. The CertainTeed building supply outlet was just a few blocks behind her house.

Running into the back yard, the woman stared up in horror. Two wide, curling black ropes were undulating like huge snakes up into the clouds. Suddenly, they converged and formed a monstrous, roaring funnel that filled the sky. As buildings under the tornado's half-mile-wide base began exploding and disappearing, Louise Covington ran back inside to open windows and unplug the television. Following a prescribed drill known to every north Texan, her husband, Shorty, hurried into the bathroom and lay down. Their teen-age son, Tifton, followed, clutching the family's frantic little dog in his arms. Before joining the family, Louise Covington glanced outside one last time. A refrigerator was cartwheeling down the street, and a neighbor's roof was coming apart. Slamming the hallway doors shut, she hurried into the bathroom.

Although the experts warn people to stay out of their cars when tornadoes threaten, many people fled anyway. A few blocks from the Covingtons, Lawana McKenzie was arguing with her 18-year-old son, Wesley, who was insisting that they flee in the family pickup truck. So persistent was Wesley, Mrs. McKenzie gave in. But they ran into a traffic jam on the outskirts of town, where cars of panic-stricken residents were backed up bumper to bumper. Seeing that the incoming lane was empty, Wes jumped the truck over the median strip and raced down the wrong side of the road, his mother screaming: "I'm not afraid of the tornado, but you're going to kill us in this pickup!" Seeing them, other drivers pulled out and followed. The McKenzies were lucky. Although they had violated a basic safety rule, they probably would have been killed had they stayed behind, for the twister wiped out their house.

Unfortunately two of their neighbors, Dennis and Grace Thorp, were not so lucky. Fleeing, too, they turned in the opposite direction to escape the same traffic jam, drove right into the path of the tornado—and died.

Back in Faith Village, Jack and Jean Russell were sprawled in the center hallway of their house with their son, David. "It sounds like a locomotive," said Mrs. Russell as a rumble shook the house and roared increasingly louder. Her ears popping as air was sucked out of the house, Mrs. Russell felt as if the house were being attacked by a giant wielding a wrecking bar. The noise was now deafening. Jack Russell was conscious of windows blowing out and things falling. He knew when the roof went because his eyes were open. When the vortex, or eye, of the tornado passed over, there was an abrupt, disconcerting calm, and just as they began to wonder if the storm were over the noise and wind came again.

By the time the vortex had passed, the Covingtons were lying with their heads next to the bathtub. The renewed force of the tornado threw Louise against the tub, stunning her. When she opened her eyes again a few seconds later, the house was still being battered. She could see the bathroom ceiling slamming up and down. "Shorty!" she yelled, rising. "We've got to go into the hall! The bathroom's fixing to go!" But their son, Tifton, sat motionless with shock, the little dog still in his arms. Louise picked up both boy and dog, lugging them into the hall. But just as she laid them down, the tornado sucked out the windows of the adjacent bedroom and the bedroom door blew inward. Horrified, not believing what she was seeing, Louise watched as the struggling dog was pulled out of the boy's arms, sucked across the bedroom, out the top of the window, and up into the dark sky. Summoning all her strength, she pulled the bedroom door shut and lay down again in the hallway.

Finally, when the tornado had passed, the Covington family began pushing away the debris that had fallen all around them— tree limbs, bricks, a refrigerator motor, hub caps. The living room was filled as well. Outside, the suburb of Faith Village lay in twisted, matchstick rubble.

Picking Up the Pieces

Almost all of southwest Wichita Falls was destroyed that day, April 10, 1979. Three funnels had come together to form one monster. The tornado had blasted a path one-half mile wide and nine miles long. Two thousand homes and hundreds of stores and other buildings were demolished.

Yet the loss was a great deal more than that. Not only was the town scarred physically but entire lifetime accumulations of personal possessions were gone irretrievably. Collections, ranging from stamps to guns, some of them rare, many the result of years of patient acquisition, were wiped out. Wedding presents, some dating back a half century. A child's favorite toy. The only remaining photograph of a dead loved one. These were the kinds of belongings that had been snatched away by the whirlwind.

And there were larger, more essential things. Entire dwellings, cars, silver and china, and—more devastating—records of savings and investments, years of mortgage payments in a home, small hoards of paper tucked away in desk drawers or files, all had disappeared. One resident of Faith Village recalls how the sudden confrontation with total loss hit one neighbor: "He was right there in his hall, and when he looked up, everything was gone. I saw him, and I'll never forget. He was ashen white two weeks after that thing." Incomes were taken away from many families, as wage earners lay injured in hospitals. The breadwinners in some households were dead.

Even pets were not spared. The Covingtons found their dog, covered with mud but unhurt physically, cowering on a couch that had been blown into a garage around the corner. She refused to go back into the house, and the family had to board her in a kennel for several weeks. When the Covingtons did get the dog back into the house it took months of luring and reassuring before the dog would set foot again in that bedroom.

The psychological toll on the human survivors was enormous. "I don't think I'll ever be the same," says Shorty Covington. "It does something to your nerves. When the weather gets that way I really get nervous, and everywhere I go I'm searching for a place for shelter. It's in the back of my mind, where can I get under this building or in that culvert, or whatever." His voice catching, the tension showing again on his weathered face, he adds: "To me, it's just like being on the front line of battle."

The Covingtons were typical of many Wichita Falls families. "I was really concerned about him," Mrs. Covington says of her husband in the days immediately following the tornado. "He couldn't make a decision. The doctor said he was in a state of shock." She suffered, too. "I broke out. I had to go to the doctor every day for two weeks."

Frequently the worst blow was felt by the children. "Our kids, one wasn't even home," says W. O. "Tiny" McKenzie. "I think it hurt him more than it did us. When he came home, everything he had was gone. It's kind of sad. All of the things you've grown up with are gone—not there anymore."

The Lesson of Wichita Falls

The tragedy of Wichita Falls and Faith Village exemplifies the wrenching loss when a family is stripped of its belongings. And that example doesn't apply solely to physical possessions. Entire nest eggs went up that afternoon. Retirement hopes and dreams were destroyed. So the loss can be not merely traumatic to the psyches of the victims but downright disastrous to every goal and ambition that people work all their lives to attain.

There are simple, sensible ways to avoid losses, or at the least—when hit by unavoidable disasters—to soften the financial impact. Take such preventive steps as checking out and eliminating fire and safety hazards in the home, keeping the family cars in good mechanical order, practicing safe driving habits, and depositing valuables in a safe deposit box to keep them out of the hands of burglars. If you should buy a vacation retreat on a mountain stream, make sure it's not in a flood plain or susceptible to landslides. All these steps and others like them will reduce the risks of loss immensely. You also can limit the size of losses when they do occur by installing such devices as smoke detectors and burglar alarms.

You can limit the totality of loss by not putting all your nest eggs in one major investment, such as a house or a single stock. When people are starting out or when they make little money, they have little choice. Their life savings are often their homes.

Nevertheless, whether one nest egg or many, there are ways for everybody to harbor his or her possessions from total loss. And that's done by using one of the oldest protections known to man—insurance. You can insure everything from your jewelry to your home to your income, and even your life itself.

Insurance plays a bigger role in the protection of your possessions than anything else. And for that reason you need to give it more than passing attention. For one thing, everybody is usually covered against some disaster or set of losses. But you need to make sure that you're not misspending your premium dollars, that you're protected where you really need it and aren't wasting money on coverage that more than likely you'll never need. And, you should be sure that your insurance is up to date and can cover the full cost of replacement.

Of course, insurance won't cover all losses. All of us are threatened every day by erosions of our investments and even outright losses because of inflation and recessions. While inflation might increase the value of your house, its impact on your savings account can be staggering over a period of 10 or 20 years. In the same way, your investments can suffer according to the whims of interest rates and the stock market. And with recessions, well, that's an even grimmer and—in today's economy—a far more ominous story.

CONSUMER TIPS

Expect the Unexpected

Protecting what's yours means being prepared for the day when everything disappears in a cloud. The ''cloud'' could be a tornado, a fire, a stock market decline, an extended illness, an auto accident or any of a host of unpleasant possibilities.

- ☐ Don't put all your nest eggs in one major investment.
- ☐ Do read on to find out what steps you should take to guard against the possibilities of loss. Consider protection for all your assets including your home, car, financial nest-egg, and even your life and health.
- ☐ Don't delay. Go ahead! Do it! *You* are the only person who can take action toward protecting what's yours. And if you don't, nobody will.

Control Your Losses

- ☐ Keep your car in good condition and be a safe driver.
- ☐ Place insurance policies, stock certificates, the deed to your home and other important papers in a safe deposit box in your bank.

Limit Your Losses

- ☐ Get rid of fire and safety hazards and install smoke detectors in your home. Protect the greatest asset you have—your life and the lives of your family.
- ☐ Take along a first-aid kit when you travel. You just may be able to lessen the severity of an injury in the event of an accident.
- ☐ Purchase and use insurance as a means of financial protection against large losses—the kind that would be difficult to handle on your own.

CHAPTER 2

KEEPING A ROOF OVER YOUR HEAD

A man in South Carolina once ran an illicit still in his home. He kept it hidden in a closet, and presumably he ran off a goodly and quite profitable amount of white lightning on the apparatus. But then one day the house caught fire, and the firemen discovered nine 55-gallon drums of mash and 22 half-gallon jugs of moonshine. Naturally the house was a sizable investment for the man and represented a major chunk of his worldly goods, and he had therefore insured it against fire. But when he filed his claim with the insurance company, the insurer balked. Stills are inclined to explode and catch fire at times, and the company complained that the man had never told them this unusual hazard existed, and that this insurance coverage therefore was void. The man sued, but the State Supreme Court upheld the insurance company, declaring: "The use to which the insured put their dwelling was so foreign to the normal uses of a dwelling as to become beyond the contemplation of the insurer."

You might not contemplate setting up a still in your closet, but you could easily lose everything you have invested in your home if you don't understand what kind of insurance you're buying and make certain you're properly covered. Too many homeowners discover too late that they aren't covered the way they expected or they have insufficient coverage. And this causes many to feel that the insurance company has treated them unfairly. Yet a major cause of such problems—and they happen every day—is lack of knowledge on the part of the policyholder. In short, you must understand what you're buying when you take out coverage on your house, or anything else, and you must make certain you have the right kind of coverage.

Occasionally some critic charges that the insurance industry employs too many people, that there are too many mouths being fed with your premium dollars. But one main reason for this is that you usually buy your insurance through an agent or company representative who gets a part of your premium dollar in return for providing you advice and counsel that will help you make certain you are spending your dollars wisely. So, get everything you're paying for. Make use of your agent, insist that he gives you that good advice—and listen to him. If you're not content with what he says, seek out another agent and see what he suggests. When protecting your house or your other possessions, tapping your agent's expert knowledge is crucial to your own well-being.

The homeowners policies that most people buy today are quite broad. (To know more about the different types of policies offered, *see Appendix B.*) But you may need supplementary coverage or higher limits to meet your own particular needs. Most policies cover:

- Your home and other structures on your property.
- Your personal property (with a few exceptions including your car).
- Your liability to others.
- Numerous other kinds of incidental, or secondary, losses.

There are separate policies for people who rent and for those who own condominiums or mobile homes. For farmers, there are special packages combining both personal and business coverages. Separate policies also are available, but not as widely sold, to cover only your liability to others or just your losses from fire, from theft or from other perils.

All homeowners policies provide for deductibles, which means that you have to assume at least a portion of almost any loss involving either your home or your other possessions.

If you shop around for insurance as a homeowner or renter, you'll find you have a wide choice of companies to deal with. While the various policy forms are pretty well standardized, companies often add special features of their own as inducements to prospective policyholders. Sometimes, variations are mandated by state laws. Generally, however, here is what you get in most of the packaged home insurance policies sold in the United States today.

Coverage on Your House

If you own your home, your house and any structures attached to it are covered—up to the dollar limit you specify when you buy the policy. Other structures on your property are covered for 10 percent of the amount of coverage on the house. That applies to any building not attached to the house, be it a detached garage, a shed or a play house for the children. And a structure is still considered detached even if it's "connected" to the house by something like a fence.

If you have other residential properties, such as a second home or a cottage at the shore, you'll have to insure them separately, even though your homeowners policy does provide limited coverage on their contents.

Since the insurance on your home is not designed to cover business activities, you can jeopardize your coverage by conducting any sort of commercial venture there. Your homeowners policy contemplates that you're going to use your dwelling "primarily as a private residence," although you'd probably get by with minding the neighbors' children now and then for a fee or with using your den to write a book in your spare time.

More elaborate business operations may require special coverage, and you can get that by adding an inexpensive endorsement to your homeowners policy. The endorsement is intended for professional people, such as doctors or lawyers, who have their offices in their homes, or for anyone who operates a private school or a studio at home. If you conduct such activities in your house, talk to your agent about the endorsement. The loss to you in case the office burns could be devastating without proper coverage.

For example, there was the case of the Virginia resident who operated a recording studio in his home, where he had about $100,000 of equipment for business purposes. He should have taken out a commercial policy.

Any business use of your home—even to give dancing or music lessons—could affect not only the insurance on your dwelling, but the coverage of your personal belongings and your liability to others. If you plan to use your home for any such purpose, the sensible approach is to talk with your insurance representative about it in advance to make certain you won't forfeit any of your insurance protection. Don't forget; they're your dollars. Make sure they buy the right protection.

Agents have saved other homeowners both money and agony with this single service. A Colorado woman made ceramics as a hobby, which turned into a business when she started selling them out of her home. Before she even had a claim, her agent reminded her that she needed a commercial policy to be covered. The same agent obtained commercial coverage for another woman who ran a day-care center in her home.

Another agent in Nebraska performed a similar service for a man who was operating an upholstery shop in his attached two-car garage.

You'll invalidate the insurance on any separate buildings if you use them, even in part, for business purposes. Even if you use a separate building to store merchandise that you sell, you'd better discuss it with your agent. The same is true of a separate building that you rent or lease to anyone other than a tenant of your dwelling for any purpose other than as a private garage. Yet it's easy to amend your policy to cover a separate structure, even an apartment over the garage that you rent out to a tenant.

Coverage on Your Personal Property

Your homeowners insurance covers your personal property anywhere in the world. So if your luggage is stolen from your hotel room while you and the family are on a trip, your insurance company will reimburse you up to the limit for which it's covered. Your personal property includes your home furnishings, clothing, tools and almost anything else people customarily have in and around their homes. (If you live in certain high-crime areas, such as metropolitan New York, you may have to pay an extra premium for away-from-home theft coverage.)

You can arrange to have your policy apply not only to your own personal property, but also to that of:

- Other people while their property is in your home (such as a microwave oven or a videotape cassette recorder that friends leave with you for safekeeping while they take off on a two-month jaunt to South America), or
- A guest or a domestic employee, whether in your home or at a secondary residence, as long as you also are staying there.

Among the few things that are not covered are pets (the policy explicitly excludes animals, birds and fish), motor vehicles and aircraft, sound recording and producing equipment you use in your car and anything you use with that equipment (such as tapes, records and discs) while it's in the car.

Beware! There are some limits to this protection. Don't allow yourself to be lulled into a false sense of security. For instance, there's a limit on what your insurer will pay under the terms of the policy to cover thefts of certain items, including jewelry, furs and other valuables, or any kind of loss of certain other items. Also, the maximum amount you can collect on other personal items may depend on where they were when damaged or stolen.

Under most policies written today, your personal property is covered—almost anywhere—for 50 percent of the amount of coverage on the dwelling. If you're a renter, you determine the amount of insurance that you need, subject to a $4,000 minimum.

Some policies cut the maximum coverage to 10 percent of the dwelling coverage or $1,000, whichever is greater, for personal property that's damaged or stolen anywhere except on your premises. That doesn't mean you'll collect any less for the loss of any single item; it's merely a cap on the total amount for which your insurer is responsible as the result of a single incident.

There's a similar limit on the coverage of furniture and other belongings that you may keep regularly at a secondary residence, such as that cottage at the shore, presumably on the theory you're not around enough to keep an eye on the place. (As a matter of fact, thefts aren't covered unless you or someone in your immediate family is actually staying there.)

Similarly, you get no personal property coverage for the furnishings of an apartment that you rent out or make available for rent on a regular basis, such as a cottage you use for one month out of the year and rent to others the rest of the time. On the other hand, your coverage wouldn't be affected if you rented the apartment only occasionally—for instance, to accommodate a friend for a couple of weeks.

The homeowners coverage doesn't extend to the property of your tenants, who are expected to insure their own things. But roomers or boarders who are related to you are protected by your insurance.

When you rent an item of personal property to others, they assume responsibility for its safekeeping, and your insurance won't cover any loss or damage while they have it. You can lend your snow blower to a neighbor down the street, but if he pays for its use you'd better hope he can pay for the repairs if he runs over it with his car.

What about other business-related personal property? Generally, if you conduct some kind of business in your home, the business-related furnishings and other equipment aren't covered under your homeowners policy. But the policy can be endorsed for a small added premium to cover such items.

If you bring business materials home to work in the evening or over the weekend, they're not covered while you're en route but your policy will cover them while they're in your home. If you take your own property to work, such as a picture to hang on your office wall, your insurer probably will respect the fact that it's yours and cover any damage that might befall it.

Check the insurance ramifications carefully when you put your household furnishings into the hands of a professional mover or into a storage warehouse. The insurance you customarily carry may well be adequate, but if you're uncomfortable about it, consult your insurance representative. Moving and storage companies usually have liability insurance, but it's intended to protect their own interest, not yours. By all means, don't lose the itemized list of your possessions that a professional mover or warehouseman will give you when he takes custody.

If you decide that the amount of coverage provided under any one or more parts of the homeowners package isn't enough to meet your need, you generally can arrange for a higher amount.

Loss of Use

Another important feature of the homeowners policy is that it will pay any additional living expenses (over and above normal expenditures) if your home is badly damaged by fire or some other insured peril and you and your family have to live elsewhere until repairs are made. Under most policies, this coverage will meet your extra expenses for up to 20 percent of the coverage you carry on your home (or, if you rent, 20 percent of the coverage on your personal property).

This protection can be a financial lifesaver. Assume you have a mortgage payment of $600 a month and a fire forces your family out of the house for a month. Since you still have to pay the mortgage, any payment you have to make for temporary quarters in a motel or an apartment can be a heavy burden. But it will be paid by your insurance company. The insurer also will pay the difference between what you normally spend for such things as food, laundry and telephone calls and what those same items cost you while you're living away from home. But one warning. The insurance company expects you to maintain your normal standard of living—not exceed it. So, don't live it up at the Ritz and expect to have all your bills paid. Obviously, it also will be up to you to keep a careful accounting of expenses to support your claim.

If you rent a portion of your home to others and it burns, your insurer will reimburse you for whatever fair rental income you may lose while the house is not habitable.

Your insurer also will pay for additional living expense and loss of rental income if civil authorities order you to evacuate your house because of damage to neighboring premises by some peril that's covered by your policy. But this provision has a two-week time limit.

Coverages Are Cumulative

Your available dollar protection for losses to other structures on your property or your other personal property, or for additional living expenses, all are in addition to the amount of insurance on the house itself. For example, if your house is insured to its full replacement value of $80,000 and both the house and your detached garage are destroyed by fire, you will receive up to the following amounts under most policies:

$80,000 for the loss of the house;

$40,000 for the loss of the contents;

$8,000 for the garage;

$16,000 for additional living expenses.

That's a total of $144,000, not counting any extras such as trees and shrubs that may have perished in the flames, which would be covered as well under some policies.

Additional Property Coverage

There are some added benefits from homeowners policies that most people are not aware of. For example, how many credit cards do you carry? Two? Six? A dozen? Probably enough so that a thief could have a field day shopping with them until you could put a stop on unauthorized bills. While your legal liability is limited to $50 per card in such a case, you could easily be bilked out of $500 if 10 of your credit cards should fall into the wrong hands. Your homeowners or renter's policy protects you against that kind of a loss up to $500. The same limit applies if you lose money because someone forges your name to a check or other negotiable instrument, or because of your good-faith acceptance of counterfeit United States or Canadian currency. And some companies have added fund transfer cards to that list.

Incidentally—and this can mean a lot—those are among the few types of losses where your deductible doesn't apply.

Other expenses that your policy usually takes care of if your home is damaged or destroyed include:

- Debris removal. Your insurer will pay the reasonable expense of hauling damaged property away. Under a new feature introduced on a limited basis in 1982, some companies will pay up to $500 to remove a tree that has damaged your property, providing that a covered peril caused it to fall.
- Reasonable repairs. If temporary repairs are necessary to protect your property from further damage, your insurer will expect you to make them. But the company will pay the expense.
- Fire department service charges. If you live in an area where property owners must pay for fire department services, your insurer will reimburse you up to $250.
- Trees, shrubs and plants. Damage to these products of nature (or your local nursery), as well as to your lawn, is covered against some perils. You stand to collect up to 5 percent of the amount of insurance on the house (but usually no more than $500 for any one tree, shrub or plant). Of course, if you grow these things for business purposes all bets are off. You'll need separate coverage. There's one exception. Since the risk is so great, trees and plants that are damaged by a severe storm generally are not covered. If a hurricane or a tornado blows a tree against the house, the damage to the house is covered—but not the tree.
- Property removal. This covers your furnishings and other belongings against any kind of damage if you have to remove them from the premises because the house is on fire or they are threatened by some other peril covered by your policy.

What's Not Covered

Some things are not covered by standard homeowners policies.

Probably the least understood of these common exclusions is flooding. Whenever a spring storm sends some creek over its banks, complaints from nearby homeowners are almost sure to follow. They usually are irate not so much because they were flooded, but because they suddenly learned that their homeowners policy would not pay a cent toward cleaning up the mess and restoring their watered-down property.

Private insurers generally don't insure homes and other fixed properties against flooding and mudslides because that kind of coverage doesn't lend itself adequately to the risk-sharing principle. For those living in flood-prone areas, the cost of insuring would be prohibitive. Property owners in areas not subject to flooding would have no need for the coverage. To offer flood insurance only to people who are the most likely to experience losses would not be practical or economical.

To fill the gap, the federal government offers flood insurance to property owners in flood-prone areas provided land-use requirements are met. (*See Page 55.*)

The water damage exclusion in your homeowners policy also applies to such things as tidal waves, water that backs up through sewers or drains and any water that seeps or leaks up from below the surface of the ground. However, the policy will cover you for damage from burst pipes and water tanks in the house, and overflows from sinks or bathtubs.

One householder found out the hard way exactly what was covered. When his hot water tank started spewing water all over his kitchen and into the hallway, he didn't know how to get rid of it. With the water lapping around his ankles, he thought of a solution. Getting a drill, he bored dozens of holes through the floor so the water could drain out. The insurance company paid for the water damage, but not for repairing the sieve effect in the floor.

Water is not the only troublesome hazard. Earthquakes are another of nature's perversities. Although they hit most parts of the United States, they have caused severe damage mainly on the West Coast. So, rather than including it in everybody's policy, insurers offer earthquake coverage separately or as an add-on to the homeowners policies of those who want it. If you live in an area that's susceptible to severe earthquakes, the coverage probably is well worth the few extra dollars it will cost you.

Other hazards your homeowners policy doesn't cover include:

- Any interruption of power or utility service that originates off your property—the kind of disruption that occurs when a power line is disabled in an electrical storm. Should that lead to a prolonged "blackout," causing the food in your refrigerator to spoil, your insurer would not be obligated to replace the loss. You'd have to fight that claim out with the utility company.

- Your failure to take whatever reasonable steps are necessary to protect endangered property and to safeguard it even after the danger has passed.
- War, including insurrection, rebellion or revolution. But riots and mere "civil commotion" are covered.
- Nuclear reaction, radiation or radioactive contamination (unless it sets fire to your property, in which case the fire is covered). Damage from nuclear accidents is covered by a special insurance program subscribed to by electric utilities that operate nuclear plants and provided by private insurers.
- The enforcement of an ordinance or law regulating the construction, repair or demolition of your home or another building on the property. Assume, for example, that you own an old house that doesn't measure up to today's building or safety code, and a local ordinance forbids its reconstruction. If the building were to be damaged substantially by fire, your insurer would pay for the fire loss but not for tearing the house all the way down.

Insurance for the Mobile Home Owner

The ranks of mobile home owners are growing fast. The Census Bureau reported an 84 percent increase from 1970 to 1980. And many insurers offer special coverage that provides basically the same insurance as the standard homeowners policy.

Other companies offer policies whose all-risk property protection even covers earthquakes and floods. Because of the high vulnerability of mobile homes to severe windstorms, insurance doesn't cover wind damage in some coastal areas that are especially susceptible to hurricanes. But you can get that coverage separately, through beach and windstorm plans operated by the private insurance industry in several states (*see Page 60*).

Many states or communities have tie-down regulations requiring that mobile homes be firmly anchored in place. Insurers often impose the same, and sometimes even stricter, requirements as a condition if you want to be covered. Always obey the public tie-down regulations or you can lose your coverage.

A distinctive feature of mobile home policies is that they cover parts, equipment and accessories—even the furniture—that you buy with the unit. While it's in your mobile home, your other personal property usually is covered for 50 percent of the amount of insurance on the dwelling.

Generally, other property coverages are similar to those offered in the standard homeowners policy, although additional living expenses are reimbursed on a dollar, rather than a percentage, basis. The allowance, subject to a specified time limit, usually ranges from $10 to $15 a day. And if you feel you need the added protection, you can increase these various coverages.

If you plan to move your mobile home, you'd better see your insurance agent first. The insurance on your mobile home doesn't travel with you, because of the additional hazards to which the unit is exposed while it's on the highways. But you can—and should—buy special coverage any time you decide to move a mobile home from one place to another.

If a mobile home must be moved to protect it, say, to get it out of the path of a threatening hurricane or forest fire—the insurance company will contribute a flat dollar amount toward the cost of the move.

As you can see, you can get homeowners insurance for just about any kind of shelter you want to call home—with the possible exception of a treehouse. And you could probably insure that, too, providing it met minimum standards of construction and tie-down. In fact, a Florida agent was even able to arrange coverage for houses built on stilts in the middle of Biscayne Bay.

CONSUMER TIPS

Home, Sweet Home

For most of us, our home represents our most important investment. In addition to the financial significance of the dwelling itself, a home contains possessions accumulated by its inhabitants throughout lifetimes or over generations.

Your home, therefore, should be a central focus when you consider **How to Protect What's Yours**.

IS YOUR HOMEOWNERS POLICY ADEQUATE?

Remember, there are several different kinds of policies designed to suit your needs as an individual.

Keeping a Roof Over Your Head 25

A homeowners insurance policy is a "package" policy which covers:
☐ Your property
☐ Your liability to others
☐ Other incidental losses such as extra living expenses you might incur if your house is damaged and you have to live in a hotel and eat in restaurants while it's being repaired.

SHOPPING AROUND
☐ When you select homeowners coverage, remember that the broader the coverage, the higher your premium.
☐ Every homeowners policy places limits on coverage for certain items—jewelry, furs, silverware, for example. Take a close look at the homeowners policies offered by several insurance companies. Look for special features with an eye toward a bargain. If, for example, you spot one policy which provides $1,000 of coverage for your silver while another (from a reputable company for the same premium) provides only $500, you should have no difficulty deciding which of the two to choose.
☐ Economize. Weigh the specific protections you get against the extra premium you will pay, and choose the policy that gives you all the protection you need, but doesn't provide coverage you don't need or feel you can get along without.
☐ Go over all options in detail with your insurance representative before you make your decision.
☐ Select a deductible. Take the largest deductible you can afford. Generally, the larger the deductible, the lower your premium.

HOMEOWNERS COVERAGES
Here are some general notes to keep in mind about homeowners insurance:
☐ Homeowners insurance is meant to cover a private residence, not a business operation. Don't make a potentially costly mistake of assuming your business is covered because you operate it in your home. Instead, talk with your insurance representative about the special endorsement you need.
☐ Your homeowners policy covers your personal belongings at home or away from home.

HOMEOWNERS POLICIES DO **NOT** COVER
☐ Pets ☐ Musical equipment used in your car
☐ Motor vehicles ☐ Aircraft
☐ Be careful about leaving furniture or other belongings at your weekend cottage or summer house—there's a limit on coverage here, too.
☐ When you move, consult your insurance representative as to whether your belongings are covered while in transit. Don't lose the itemized list given to you by the mover. Make and **keep** an extra copy to be safe.

SOME MAJOR EVENTS **NOT** COVERED BY HOMEOWNERS INSURANCE:
☐ Flooding / mudslides
☐ Earthquakes
☐ Nuclear accidents
Ask your agent about coverage for these unusual occurrences.

CHAPTER 3

GETTING FULL PROTECTION WITHOUT OVER- SPENDING

Many homeowners tend to equate the amount of insurance they need on their dwelling with the market value of the property. But if you do that you may be buying more insurance than you need. You don't buy a homeowners policy to cover damage to your sidewalks or the land around your house. Your main concern is protecting your investment in the house itself, any other buildings on the property and your personal belongings. The land is going to remain intact, and probably retain its value, even if the house and everything in it is destroyed.

Thus, you should find out how much it would cost to replace the house. If the estimated market value of your property (house and land) is $100,000, and the projected cost of replacing the dwelling (rebuilding from scratch) is $80,000, then $80,000 is the figure you should work with for insurance purposes, not $100,000.

That's in case of total loss. But you may have gathered the information (true) that most homes that catch fire, or even are struck by a tornado, escape with only partial damage. Therefore, you may wonder if you couldn't take a chance and carry less coverage. For instance, if you figure the chances are slim that you'll take more than a 50 percent loss, you might ask if you shouldn't insure for just half the cost of total rebuilding. don't do it. Under the terms of a typical homeowners policy, you wouldn't be adequately protected.

Here's why:

1. If you should have a total loss, you'd collect no more than the face value of your policy. That much is obvious.
2. The amount of insurance you carry on the house determines automatically (unless you have other arrangements written into your policy) the amounts you can collect for other buildings on your property, your personal belongings, and additional living expenses you might have while your house is being repaired. These amounts are usually figured as a percentage of the primary coverage on the house. If you had only partial coverage on the house, the amounts you could recover on these other items would be less.
3. In order to collect the full replacement cost for even partial damage to your house, you have to have coverage in an amount equal to at least 80 percent of its full replacement value. These are the standard terms of most insurance policies on homes.

This last point is a little confusing to many people. The traditional way to pay insurance losses is to determine the replacement cost of the property damaged and then deduct an amount for depreciation. For example, if a roof with a 20-year life is destroyed after 10 years, the insurance payment would be depreciated by 50%, or half the cost of a new roof. You wouldn't be eligible to get a new roof for one that is half worn out. But under the replacement cost provision, insurance companies offer a bonus if you insure to 80% of the replacement cost—in effect they pay to replace new for old without taking depreciation into consideration. Here's how it works:

Assume the replacement value, or rebuilding cost, of your home is $80,000. Coverage for $64,000 (80 percent of $80,000) meets the "80 percent of replacement value" requirement. If a fire should cause $10,000 damage to your living room, your insurance company would pay you that amount, less whatever deductible you have chosen, to cover the necessary repairs.

But if you decide to gamble on $48,000 worth of insurance in order to lower your premium, you're insured for 60 percent of your home's replacement value. In turn, that equals only 75 percent of the minimum amount ($64,000) that the terms of your policy allow you to carry and still collect the full replacement cost. Therefore, you can collect only 75 percent of your $10,000 repair cost ($7,500)—or, at the very least, actual cash value (replacement cost less depreciation). In either case, you won't collect $10,000. And you'll be out of pocket for the difference. Partial losses to any other buildings on your property would be protected the same way.

So, it's in your interest to insure for at least 80 percent of the replacement value of your house. Many insurance companies today won't even deal with applicants who want less than 80 percent coverage, and some companies may even require full, 100 percent, coverage. It saves them a lot of disappointed customers who didn't understand what they were buying.

Obviously, to determine how much insurance you need it's necessary that you have a good "fix" on what it would cost to replace, or rebuild, your home. A good starting place for getting the information is your insurance agent or company representative. Many insurance companies have produced guides which enable you to estimate the replacement cost of your home yourself. You fill in the square footage or number of rooms, together with other factors such as type of construction, and multiply by a number reflecting current construction costs in your locality. Generally, costs tend to be lowest in the Deep South, highest in Alaska and in between in other areas of the country—reflecting differences in labor costs and availability of materials.

If you want to be sure about it, you can retain for a fee the services of a professional appraiser. This is probably worth the

Getting Full Protection Without Overspending

expense, and it is a particularly good idea if your home is built like a mansion, or is otherwise unusual in its construction.

A mistake some people make, especially when they first buy a home, is purchasing just enough insurance to cover the amount of their mortgage. If you're among them, you should increase your coverage before you lose your equity to a fire. Otherwise, all you're doing is protecting the bank or savings and loan that holds your mortgage. And the money you've put into the property will all go up in smoke.

One final note: If you add a room or make other substantial improvements, the value of your property should increase. Be sure to change your insurance coverage to protect your added investment.

Guarding Against Inflation

But don't change your coverage only if you make an improvement. Take a fresh look at it every year. Many thousands of homeowners are underinsured merely because they don't take the trouble to keep their coverage up to date. Inflation might have increased the market value of your home. It also can double or triple the cost of replacing it. For example, from 1971 to 1981 the average cost of repairing a home rose 135 percent. That means that a house built for $30,000 in 1971 would have cost $79,500 to replace in 1981.

Even if inflation has grown a little less rampantly in your area, costs haven't stood still. And unless you confer with your agent or representative at least once a year to make certain that you still have adequate coverage, a fire or other catastrophic loss could leave you in a serious bind.

Some policies deal automatically with that problem. They have escalation clauses tied directly to changes in local residential construction costs. If construction costs go up while your policy is in force, coverage for your dwelling, your personal property and additional living expenses will be adjusted upward at the same rate.

But if you don't have such a clause in your policy, you might find it worthwhile to ask your insurance representative about

Standard Amounts of Coverage Under The Special Form Homeowners Policy

(Assuming a house with an $80,000 replacement value insured to full value, 80% of value and 60% of value)

Property Coverages	Insured to Replacement Value	Insured to 80% of Replacement Value	Insured to 60% of Replacement Value
Dwelling	$80,000	$64,000	$48,000
Other Structures on Property	8,000	6,400	4,800
Unscheduled Personal Property	40,000	32,000	24,000
Additional Living Expenses	16,000	12,800	9,600
Trees, Shrubs and Plants	4,000*	3,200*	2,400*
Liability Coverages**			
Personal Liability (for each occurrence)	$25,000	$25,000	$25,000
Medical Payments to Others (per person, regardless of fault)	500	500	500
Damage to Property of Others (per occurrence, regardless of fault	250	250	250

* Subject to a maximum of $500 per item.
** Some companies have increased, in some areas, the amounts of coverage for personal liability, medical payments to others and damage to property to others.

---

the "inflation guard endorsement" which many companies offer. It automatically increases your policy limits each quarter by a fixed percentage.

The Price You Pay

While the cost of home insurance generally is subject to such economic influences as changing costs of materials and labor, it also is affected by frequency and size of claims in your area. Home insurers take other considerations into account as well, such as the efficiency of your local fire department; the dependability of the city's fire alarm system; the nearness and adequacy of a water supply, and the ready availability of emergency personnel.

Once the rates are established for your community, you in effect set your own premium on the basis of what you have to insure, how much you insure it for, and the kind of policy you choose.

The table below provides an example of how homeowners premiums vary by community, by type of construction and by the amount of coverage you may select. The prices all are those of the same large insurer, for a typical homeowners policy with a $100 deductible, as applied to each of two types of home in July 1982. In each case, the dwelling is assumed to have a replacement value of $80,000.

While costs may vary according to the kind of building you own or where you live, they also can vary for a similar house in the same area. Some companies might offer more coverage for certain types—such as personal possession limits—than their competitors. That means it always pays to shop around. Ask your agent what choices you have. If you aren't satisfied with his offerings ask several agents. But do look at all the choices.

Ways to Save

There are a number of ways you can lower the cost of insuring your home. If you'll look back at that last table, for example,

Typical Costs of Insuring an $80,000 Home*
(Assuming an HO-3 homeowners policy with a $100 deductible)

Locality	Insured for $64,000 (80% of Value)		Insured for $80,000 (Full Replacement Value)	
	Frame	Masonry	Frame	Masonry
Jacksonville, Fla.	$310	$245	$348	$277
Wethersfield, Conn.	329	260	359	286
Suburban Lincoln, Neb.	435	345	471	371
Sacramento, Calif.	474	393	486	402

* Premiums shown are for homes located either within 1,000 feet of a public fire hydrant or within three miles of a fire department. The premiums would be higher if neither of those conditions were met.

you'll find that you can knock off a significant chunk of your premium by insuring your home to 80 percent of its replacement value instead of for full coverage.

Another approach that many home owners find more palatable is to manage their insurance in such a way that they have full protection against the potentially catastrophic losses while assuming more of the smaller losses themselves. In other words, by increasing the size of their deductibles. Take the family with the $80,000 frame house in suburban Lincoln, Neb. As the table shows, they would pay $471 for a full-coverage policy with a $100 deductible. For the same policy, they'd save $47, or about 10 percent, by switching to a $250 deductible, or $94 (about 20 percent) by selecting a $500 deductible. Although they would have to assume the expense of any loss up to the deductible amount, they still would be completely protected against the rest of a major loss.

Take, for example, the case of the householder who was awakened from a sound sleep one balmy summer night by the buzzing of an insect in the bedroom. Hoping to rid the room of the intruder, he reached for an aerosol can of insect spray, leaped out of bed and began blasting away, pursuing the sound of the buzzing. In the morning, he discovered to his dismay that the can he had grasped in the night was not the insecticide, but red spray paint. He had spray-painted walls, rugs and furnishings of the bedroom in an interesting but hardly professional, manner. Of course, he paid up to the deductible. But since the damage considerably exceeded the deductible, the insurance company took care of all the rest.

Getting a Discount

If you have a burglar alarm or a sprinkler system that meets certain standards, or even smoke detectors or dead-bolt locks, you might be able to get a discount on your insurance premium. With an alarm system, the amount of the discount usually will depend on whether the detection equipment is connected to a central alarm monitor system, linked directly to a police or a guard station, or else geared to wake the neighbors by sounding an alarm outside the house.

At least one state, Texas, requires insurers to provide a 5 percent premium credit for homeowners who install dead bolts, solid doors and window locks, as long as they are checked and approved by a certified inspector of the Texas Crime Prevention Institute.

Some insurance companies offer discounts in some states for such things as:

- Participation in a program through which you etch your driver's license number or some other form of personal identification on articles of value to enhance their recovery in case of theft;
- Equipping homes with solar energy systems (which are deemed to be less fire prone even if they have a back-up oil or gas heater);
- Building earth-sheltered homes (better protected against violent weather and more fire-resistant);
- Achievement of retirement status (on the ground that retired people—at least those who are 55 or older—are better insurance risks since they spend more time at home where they can look after their property).

Many insurers routinely provide premium credits to owners of new or relatively new homes, which often are insured to their full replacement value and which, on average, are less subject to insurance claims because of the newness of their construction, wiring and safety features. Some companies provide a discount of as much as 20 percent for a brand new house and then scale the discount down each year for the next five to nine years.

A Household Inventory

It won't reduce your premium, but it will help you collect your claim when you have a loss if you take a detailed inventory of all your possessions and list their value.

A household inventory, basically, serves two purposes:

1. It gives you a good grasp on the overall value of what you own and may warn that you need more personal property coverage than your policy provides, and
2. It will help you document your claim in case you have a loss.

Think about it. If a fire were to destroy all or part of your home tomorrow, could you rattle off a list of all your lost possessions to your insurance adjuster? Try it. Take a pencil and a sheet of paper into the kitchen and make a list of everything in the living room. Or the basement. Or a closet. Do the best you can to include the original price of each item or at least its estimated value and what it would cost to replace it. Don't overlook the contents of drawers and things that are hanging on the walls. Then check your list with what is actually there.

You'll find it isn't easy. But it will be even harder if you have to come up with a believable itemization after you've had a loss. An inventory will eliminate that problem.

You can probably get an inventory checklist from your insurance representative, or from a county extension service. Or you can use a pad of paper, preferably separate pages for each room.

Make your list as complete as possible. List where and when you acquired each item, its original cost and what it would cost to replace it now. Put down the model number, if any. And give a description. For instance, if it's a dining room table is it made of oak, pine or mahogany?

Do the best you can with gifts, hand-me-downs and heirlooms. Even if you can't say much about their origins or values, describe them.

You might want to skip that kind of detail with standard items of clothing, although you really should include a count of such things as socks, shoes, shirts and dresses. If you customarily pay $150 for a suit and you just blew $500 for a new one, put it down.

Whenever you can, attach to your inventory sales receipts or purchase contracts for major items. And, if it's practical, support your inventory with photographs, including close-ups of contents of drawers and of expensive items such as paintings, antiques, silverware and jewelry. An alternative, if you have or can rent the equipment, is to use videotape with an audio narration. There are even some service organizations now that will do this for a fee.

You may need professional appraisals of high-cost items or the kinds of things which typically appreciate in value. A jeweler to attest to the authenticity and value of your jewels, a rug

Getting Full Protection Without Overspending

merchant for your Oriental carpets. You can have the entire inventory appraised professionally for a price, but it can be expensive. One national organization said recently that its standard minimum for such a job was $500, excluding antiques, which would require still other outside, expert appraisals.

Once you've completed your inventory, don't forget to add new items of value as you acquire them and delete items you dispose of. You might want to keep a copy of the inventory itself at home for reference or for updating. But be sure to put the original, along with the sales receipts, photographs and other supporting documents, in a safe deposit box or some other safe place away from home.

CONSUMER TIPS

Going for the Most

Remember, it's your responsibility, with the help of your insurance representative, to figure how much insurance is enough for you. Get the most without overspending.
- ☐ Base your coverage on your home's replacement cost, not its market value.
- ☐ Make certain your policy covers at least 80 percent of the replacement value of your home.
- ☐ Review your home insurance policy at least once a year.
- ☐ Talk with your insurance representative about increasing your coverage if you add a room or make other substantial improvements to your home.
- ☐ Decide whether your homeowners coverage is keeping pace with inflation.
- ☐ Update or modify coverage as needed.

USE THESE MONEY-SAVING STRATEGIES:
- ☐ Save $ by insuring to 80 percent of replacement value rather than to full (100 percent) of replacement value.
- ☐ Save $ by taking the largest deductible you can afford.
- ☐ Save $ by taking all discounts or credits for which you may be eligible.

FIND OUT WHETHER YOUR INSURANCE COMPANY OFFERS DISCOUNTS FOR:
- ☐ Fire and/or burglar alarm systems
- ☐ Smoke detectors
- ☐ Sprinkler systems
- ☐ Deadbolt locks on exterior doors
- ☐ Personal identification numbers etched onto valuable articles
- ☐ Solar energy heating systems
- ☐ Earth sheltered homes
- ☐ Achievement of retirement status
- ☐ New homes, with new wiring and safety features

Keep an accurate, up-to-date inventory of your furniture and other major belongings in a safe deposit box or other safe location away from your home.

Secure special coverages for jewelry and other valuables.

CHAPTER 4

HOW TO KEEP YOUR HOME FROM BANKRUPTING YOU

As do most accidents, it happened in a matter of seconds. A 16-year-old boy was untying a dinghy at a friend's waterfront home when an old concrete dock collapsed and pinned the teen-ager into the small boat. He suffered a compound fracture of the thighbone and lesser fractures of the leg.

There was no question that the property owner was responsible for the mishap; he'd known about the poor condition of the dock for some time. His insurance company paid out just short of $15,000 to cover the youth's medical bills and lost wages (the boy was employed part-time), and added another $2,000 for "pain and suffering."

While the average property owner possesses neither a dock nor a dinghy, this boy's accident is typical of the kinds of liability your home can expose you to.

So it makes sense to check that you're fully protected from liability suits, and that even the unusual risks are provided for. All homeowners policies include liability coverages which protect the interests of any family member who lives in the house or anyone under age 21 who resides there and for whom a family member is responsible. Your policy covers not only your legal liability to others, but minor damage to other people's property whether you are legally liable or not.

For example, if your pet scratches a visitor or tears up a neighbor's chair, that's covered, too. After all, you're responsible for your pet's actions, and your guests and neighbors expect you to keep it under control, even though that's sometimes difficult. Take the homeowner whose pet was a large and vigorous mutt of uncertain ancestry, named Brutus. His next-door neighbor was the proud owner of a pedigreed poodle of championship lineage, Fifi. There was a fence between the houses, but one night Brutus, in a fit of passion, tore it down and joined Fifi for a brief moment of romance. Outraged, Fifi's owner asserted that Brutus had contaminated her pedigree's bloodline, besides destroying the fence. She lodged a claim, and her neighbor's insurer paid off under his homeowners liability coverage.

Personal Liability

A risk to which you're always exposed as a homeowner or a renter is a demand (whether it's in the form of an insurance claim or a lawsuit) that you or a member of the family fork over a sizable chunk of money to someone else because of a loss or injury attributed to your negligence.

Take another shaggy dog story, for example. A Colorado homeowner's daughter, visiting a friend, opened the back door and let her friend's Alaskan husky out. The husky ran down the block and promptly chewed up a smaller dog. Taken to the vet, the small dog suffered a heart attack and had to be operated on, after which he was hospitalized for a week. Since it was the negligence of the first homeowner's daughter in letting the husky out that precipitated the incident, her parents were presented with the small dog's $700 medical bill. Their homeowner's liability coverage paid.

Your liability insurance is designed to take care of such claims, whether they involve bodily injury, property damage or both, up to the limits of your policy. The kinds of problems that can prompt such claims are almost endless. For example:

- A visitor tripping over a child's toy.
- A passerby stumbling over a crack in the sidewalk.
- Your dog biting the mailman's leg.
- Injuring a passerby with your umbrella during a heavy rain.
- A lapse of judgment which causes a tree you're cutting to crash into a neighbor's house.

Take the unusual and frightening case of one homeowner. Faulty judgment of a different kind paved the way for a substantial claim against him when he was probably at home in bed at the time of the accident. His offense lay in permitting a guest to drive away from a party in an inebriated condition. The guest struck two pedestrians with his car, killing one and injuring the other. A claim alleging negligence of the host resulted in the payment of $150,000 in damages.

Your personal liability insurance covers accidents that happen anywhere in the world. If a claim is filed against you as the result of an injury or property damage, your insurance company will take over. If it is agreed or a determination is made that you were legally liable, the company will pay the damages up to the policy limits. In the event of a lawsuit, the company has the option of settling out of court or defending the case on your behalf. Your insurer will pay the legal costs of the defense whether or not you ultimately are held to be legally responsible. Of course, your insurance company will look to you to cooperate in its investigation and defense of any claim, and will pay any reasonable expenses up to $50 a day that you incur at its request.

Generally, the minimum liability limit in any of the homeowners policies is $25,000, although some companies have increased that minimum to as much as $100,000 because of inflation. Additional amounts of liability coverage are available at relatively little additional cost.

The recent move to raise the minimum limit makes it clear that some people are finding $25,000 of liability coverage totally inadequate. If your financial condition is healthy enough to make you a target for a sizable claim, you should consider extra coverage. A more reasonable limit is $100,000 to $300,000. If you are especially vulnerable, add "umbrella" coverage. The added fee is minuscule, but the coverage can save everything you have.

Other Liability Features

At least to some extent, most homeowners policies cover any personal liability you might assume under a contract or agreement—providing you put it in writing. Some policies cover any such contract—such as the kind you might sign if you hire a hall for an anniversary celebration and agree that you'll be responsible for any accidents that might occur during the affair. Others limit the coverage to contracts relating to the maintenance of your property, whether it's your home, an unimproved lot that needs clearing once or twice a year, a cemetery plot or even someone else's place where you are living temporarily.

An Oregon man named Kuhn, whose home was situated on a slope leading down to a small lake, had some improvements made to his property that elevated it above that of his next-door neighbor. The city owned the property at the top of the slope. Because the city failed to provide proper drainage, water from a heavy rainfall ran onto Kuhn's property and was deflected onto his neighbor's land, causing severe erosion. The neighbor filed a claim against both Kuhn and the city. The city repaired the damage, and Kuhn's insurance company paid $1,200, which was determined to be his share of the cost.

Don't forget. In no case does this liability coverage apply to any business-related contract or agreement, whether it's in writing or not.

Your insurer usually won't pay for damage that you do to your own property or to property that you rent, but insurers customarily waive that exclusion with respect to fire, smoke or explosion damage to property that you are renting or using.

For instance, you probably have insurance protection if you accidentally set fire to your motel room and cause extensive damage. The motel's fire insurer would probably pay for the damage, but could then turn to you for reimbursement which your homeowners policy would cover if you were found legally liable.

You usually are covered for any injuries to your household employees, but the coverage doesn't apply to a bodily injury to anyone to whom you provide, or are required to provide, benefits under a workers' compensation, occupational disability or occupational disease law. So, if you have workers' compensation for your household employees, don't waste money taking out other special accident coverage on them.

Medical Payments

Often, accidents cause minor injuries which are inexpensive to treat but can become relatively burdensome in both time and dollars if payment of the bills has to wait until the court decides who's legally liable. The medical payments section of your homeowners policy provides for payment of such bills without regard to fault in case of injuries to others on your premises. That means that if a friend should fracture a collarbone during a touch football game in your back yard, your insurer would pay the medical bill as long as it fell within the policy limit.

The medical payments coverage also applies to accidents that happen away from your home if they are caused by you, a member of the family, a domestic employee (if the accident happens in the course of the employment) or a pet.

The basic amount of medical coverage is $500 for each person injured, although some companies have increased that amount to $1,000. And you can buy larger amounts of coverage. But, remember, this is part of the liability coverage and it doesn't pay for injuries to you or to family members who live with you. You need other insurance for that.

Damage to Others' Property

Like the medical payments coverage, this feature enables you to discharge what you might consider a moral obligation. It pays for minor damage that you or someone in the family might cause to other people's property regardless of who is at fault, or even when no one was at fault. As an example, your insurer would pay for the damage if a guest at a party in your home were to have his clothing burned beyond repair by standing too close to a lighted candle. Damage caused intentionally by children under age 13 is covered because it is considered accidental. Some insurers will pay up to $250 for any one accident, others up to $500.

What Your Policy Doesn't Cover

Once again, the liability protection provided by homeowners policies doesn't cover bodily injury or property damage you might cause in the course of your work. That's true even if it's only a part-time or moonlighting job. Ask your agent or company about a rider on your homeowners policy to cover you if you moonlight.

This exclusion of business-related damages frequently causes costly disappointments to policyholders who didn't understand what they were buying. Also, since coverages are so complete and cross lines so frequently, it's not always easy even for the insurance company to know precisely whether you're covered under one policy or another. This sometimes causes misunderstandings between policyholders and their insurance companies. For example, a man returning home from a business trip accidentally struck and injured another person with his suitcase in the airport. A court eventually decided that even though the trip was business-related, the man's careless handling of the luggage was an action not "ordinarily incident" to his job, and his homeowners insurance paid the claim.

What constitutes a business pursuit in the first place? Obviously, if you conduct any kind of full-time business at home, whether it's a professional office, a dance studio or a repair shop, you should arrange for extra coverage to protect your assets in case of a business-related injury to someone or damage to someone's property.

One kind of business problem that is explicitly excluded from liability coverage is what is known as professional malpractice. If you're a doctor, lawyer, architect, or anyone else who provides professional services, you need special coverage against any claims that might be lodged against you should you make a mistake or fail to perform some crucial service for a client.

But what if you injure a child with whom you're baby-sitting for a fee? Or someone trips over a carelessly-placed piece of furniture at your garage sale? Not all insurers necessarily follow the same line of reasoning, but you could be in trouble if you don't have special coverage. As an executive of one major insurance company puts it: "If it's something that's done regularly for a profit, we usually consider it to be a business pursuit."

Under that interpretation, he says, an injury caused by some infrequent activity, such as a once-a-year garage sale, would be covered. So would an occasional baby-sitting job. But insurance companies would consider a regular baby-sitting job a business pursuit, and you'd need special coverage to protect your liability should you drop or otherwise injure the baby.

Remember, though, that every insurance company might not look at it that way. For your own protection, you'll be well advised to check with your insurance representative before getting involved in any sideline venture for profit, especially if there's a chance someone can get hurt.

Working Minors—You may unwittingly be exposed to an unprotected damage claim if your young son or daughter takes on a newspaper route or any of a number of odd but regular jobs—such as mowing lawns, baby-sitting or shoveling snow.

You could be in for a hard time, for example, if Tommy should accidentally knock down a pedestrian with his bicycle while flinging a newspaper onto a customer's front porch. Delivering papers on a regular basis for pay clearly falls into the category of business pursuits, and your insurance company could deny liability unless your policy provides otherwise. Some insurers dispose of this problem by inserting into the policy a provision that the business exclusion doesn't apply to occasional part-time jobs held by minors. If your policy doesn't contain some such provision, ask your insurance representative to provide you with assurance—in writing, if possible—that your liability is protected when the kids go to work.

For some jobs, such as that of newspaper carrier, your policy can be endorsed to cover you in case of an accident. The additional coverage may turn out to be well worth the small cost.

Rental Property—If you rent any residental property and there is an accident for which you're liable, it generally is treated as business-related. And usually your liability isn't covered under your regular homeowners policy. In effect, when you rent out residential property—whether a condominium, vacation cottage, house or other dwelling—and it's on a regular basis, your insurer looks upon it as a business venture.

But this exclusion doesn't apply to a house that you rent to someone only occasionally for use as residence. And your liability is still covered if you rent a portion of your house to one or two (no more) roomers or boarders or if you rent a part of the house for use as an office, school, studio or private garage. Of course, the person or group that you rent to would need their own liability coverage to protect themselves from claims.

Vehicles and Boats—As a general rule, the liability coverage in your homeowners policy doesn't cover accidents involving your motor vehicles or aircraft. There is one common exception. Your insurer will pay a legitimate claim if you operate a motorized golf cart carelessly while you're using it for a round of golf and have an accident.

Some insurers have removed model planes from the category of aircraft, which means that you'd be covered if you get one of those sophisticated toys airborne and it crashes onto the head of an innocent citizen.

If you are a boating enthusiast, your homeowners liability insurance will cover you up to a point depending on the boat. You're covered if you own or rent an inboard or inboard-outboard boat of not more than 50 horsepower or a sailboat less than 26 feet long. And you're covered if you own an outboard power boat of 25 horsepower or less. But if your boat or motor is bigger, you'd better see about special coverage before you go sailing again.

Workers' Compensation

Although workers' compensation generally is thought of as a coverage for business and industry, you might need it at home. It is available as a separate protection for homeowners and renters to cover them if their domestic employees—whether they be full-time or part-time—are hurt on the job.

At least two states, New Hampshire and New Jersey, require that every policy that includes comprehensive personal liability coverage—such as the homeowners and renter's policies—also have workers' compensation coverage for household employees. In California, the law says you need the coverage if you own a home and employ domestic workers either full-time or part-time.

Essential Protection

Everything considered, liability coverage—whether it's yours or the other party's—can be like the cavalry riding to the rescue.

Lots of problems are unforeseeable. For instance, there was the house on Long Island, N.Y., that was temporarily unoccupied and stood next door to a house that had been abandoned. The city, which had taken over the abandoned dwelling, ordered its demolition. The wrecking crew duly arrived, went to work—and wrecked the wrong house. When the city manager drove by to inspect the job, he was shocked to find the crew destroying the perfectly good house next door. Halting the work, he notified the horrified owner. But it was too late; the house was a total loss. The wreckers' liability insurer agreed to pay.

The cost of liability coverage is a small portion of your homeowners or renters insurance premium, but the protection can be essential to your financial health. So understand what coverage you have—and what you need.

CONSUMER TIPS

Are You Liable?

Liability insurance is an important part of your homeowners coverage and an integral part of protecting what's yours.

PERSONAL LIABILITY

Liability insurance is designed to protect you as a homeowner or a renter because you are always exposed to a risk of a lawsuit or insurance claim due to loss or injury attributed to your negligence. This insurance covers:

☐ Bodily injury
☐ Property damage
☐ Both

This insurance also covers accidents that occur anywhere in the world. If a claim is filed against you as the result of an injury or property damage, your insurance company will take over.

The usual amount of liability coverage under a standard homeowners policy is $25,000. Consider increasing your limit to $100,000 or $300,000 depending on your financial condition.

For more than $300,000 coverage, discuss an umbrella liability policy with your agent.

OTHER LIABILITY FEATURES

☐ Most homeowners policies cover, at least to some extent, any personal liability that you might assume under a contract or agreement—providing that it's in writing. Some policies cover any such contract. Others limit the coverage to contracts relating to the maintenance of your property.

☐ Homeowners policies do not cover any business-related contracts or agreements whether they're in writing or not.

- ☐ Your insurer usually won't pay for damage that you do to your own property or to property that you rent, but it customarily waives that exclusion with respect to fire, smoke, or explosion damage to property that you are renting or using.
- ☐ You usually are covered for any injuries to household employees for which you are liable.
- ☐ The coverage does not apply to bodily injury to anyone to whom you provide, or are required to provide, benefits under workers' compensation, occupational disability or occupational disease law.
- ☐ Another feature pays for minor damage that you or someone in your family might cause to other people's property regardless of who is at fault, or even when no one is at fault.

MEDICAL PAYMENTS
- ☐ The medical payments section of your homeowners policy provides for payment of medical bills for minor injuries to guests in your home.
- ☐ The medical payments coverage also applies to accidents that happen away from your home if they are caused by you, a member of the family, a domestic employee (if the accident happens in the course of the employment) or a pet.

LIABILITY INSURANCE DOES **NOT** COVER:
- ☐ Injuries or property damage caused in the course of your work.
- ☐ Generally, accidents that involve motor vehicles.
- ☐ Professional malpractice.
- ☐ Property you rent to others.
- ☐ Accidents that involve your motor vehicles or aircraft, as a general rule. (There is one common exception. Your insurer will pay a legitimate claim if an accident results from your careless operation of a motorized golf cart while you're using it for a round of golf.)

IMPORTANT "DO'S AND DON'TS."
- ☐ **Do** become acquainted with the liability coverages provided by your homeowners insurance.
- ☐ **Do** phone your insurance representative when you cause an accident or injury, even if you think you're not covered. It's certainly worth a phone call to discover that you may get some help in protecting what's yours.
- ☐ **Do** make an assessment of your personal financial situation. Chances are, if you have more than $25,000 worth of assets, you need to have more than that amount of liability insurance.
- ☐ **Don't** depend on your homeowners insurance to cover your liability as a businessperson.
- ☐ **Don't** make the mistaken assumption that your homeowners liability policy covers your teenager while he's working at his part-time job.
- ☐ **Do** make it a point to discuss with your agent the workers' compensation laws in your state and how they affect coverage for household employees.

CHAPTER 5

WATCHING OUT FOR YOUR PERSONAL POSSESSIONS

You return home from a weekend in the country, and you can't believe—don't want to believe—the sight that greets you when you open your front door. Contents of closets are strewn across the floor. Cabinets have been broken open. Upstairs, dresser drawers have been dumped out. The place looks as if a hurricane had swept through.

But that's not the worst. A quick check tells you your most valuable possessions are missing. The mink coat that had been hanging in the front closet. Your sterling silver flatware. Your Aunt Martha's diamond ring that you'd been meaning to put into your safe deposit box at the bank. A wallet containing a hundred dollars in cash and some valuable papers.

You've been burglarized.

Not only are you shocked and sickened by the thefts, but unless you've arranged for special insurance coverage, you may be in for a couple of other shocks, too. First, now that you have reason to add it up, you realize your lost possessions were worth a lot more than you had thought. Second, you discover that what you're going to get from your insurance company comes to no more than a fraction of their value.

When you take out a standard homeowners policy you insure many kinds of items classed as personal belongings, for only an extremely limited amount. Many companies have started raising those limits to allow for inflation, but even then the standard amount of coverage that applies to your valuables may still be woefully inadequate, certainly so far as your own perception of their worth is concerned.

Here are the most common upper limits of coverage on specific kinds of items (the amounts in parentheses are typical of new ceilings now being offered by companies in certain states):

1. $100 ($200) on money, bank notes, bullion, gold, silver, platinum, coins and medals. A single item in a valuable coin collection could be worth more than you'd recover if the entire collection were to be stolen or to perish in a fire.
2. $500 ($1,000) total on securities, accounts, deeds, evidences of debt, letters of credit, notes other than bank notes, manuscripts, passports, tickets and stamps. If you're a philatelist, this probably won't come close to compensating you adequately for that stamp collection you've been laboring over for years.
3. $500 ($1,000) on boats, including their trailers, furnishings, outboard motors and equipment.
4. $500 ($1,000) on trailers not used with watercraft.
5. $500 on grave markers. (Some companies don't impose any limit on this item.)
6. $500 ($1,000) for loss by theft of jewelry, watches, furs, precious and semi-precious stones.
7. $1,000 ($2,500) for loss by theft of silverware, silver-plated ware, goldware, gold-plated ware and pewterware.
8. $1,000 ($2,000) for loss by theft of guns and firearms (a particularly "hot" item for burglars).

In each category, the dollar limit is all you will collect each time there is a loss, no matter how many items are involved. For example, if that mink coat and the diamond ring together were worth $5,000 and your limit under category 6 were $500, you'd collect no more than $500 altogether—not $500 for each item. But you would collect a separate amount for the stolen silverware, which comes under a different category.

Some of these items—if large or unusually valuable—can be protected with special policies. Large boats are one example. You can get a policy just to cover your boat, and if it's powerful and liable to hurt people, you should. But don't make the mistake of one Michigan sailor. Boat policies are written for the specific areas that the boat normally sails in. That man's boat was insured for the Great Lakes area. But for one vacation he took the boat to Florida, where his father drove it and crashed and sank it. The man should have obtained an endorsement to his policy giving him coverage in Florida. Because he didn't, he lost his entire investment. So, if you do have a special policy that covers the possession only in a specific place or territory, don't forget to notify your agent if you move it somewhere else.

There is a way to protect your other possessions, often at a small fraction of your usual homeowners premium. For an additional premium, you can either increase the limits on the various kinds of personal items, or else buy a personal articles "floater" which is available either as a separate policy or as an endorsement to your homeowners insurance. The term "floater" derives from the fact that the coverage applies wherever the property is located at the time of loss. The coverage, in effect, "floats" with the property.

With the personal articles floater, coverage is provided on an itemized basis as well as by category. You supply a full description of each item along with its value—you'll have to support your application with a sales slip or an appraisal in most cases—and the coverage is tailored to insure each article accordingly.

The floater provides "all-risk" coverage which can be applied not only to jewelry, furs, silverware, stamps and coins, but to personal items that people commonly take with them when they travel, such as cameras, golfing equipment and musical instruments. The floater even covers fine arts and antiques.

Through the all-risk feature, you can protect yourself from such mishaps as:

- The loss of a camera during a vacation trip;
- A dropped watch;
- The spattering of paint on a valuable painting or other work of art (but not any further damage done during the restoration process);
- Someone breaking an antique chair.

In one bizarre case, a woman had left her engagement ring on the kitchen window sill while she washed dishes, and a bird, probably attracted by the flashing of the diamond in the sun, swooped down, grabbed the ring in its beak and flew off. The woman quickly ran outdoors, shouting at the "thief," but only frightened the bird into dropping the ring into the middle of a pond. The insurance company paid her claim; with a floater's itemized coverage, payment on such losses is automatic.

Unfortunately, another woman didn't recover her loss. She was wearing her engagement ring, worth several thousand dollars, on a ferry ride when it slipped off her finger into the river. If she had had the personal articles floater on her basic homeowners policy, the loss would have been paid.

Somewhere in between those two is the case of the Florida woman who reported the loss of her ring, collected its value from her insurance company, and then found the ring. Puzzled about what to do with the money—and presumably embarrassed to call her agent—she gave the money to the Red Cross. This, obviously, wasn't the "right" thing to do, even though it might have eased the woman's conscience. Her agent could have advised her that her legal as well as her moral obligation was to the insurance company, in a case where a "lost" article later turns up.

A case where a claimant didn't lose her property at all, but still collected under scheduled coverage, was that of a woman who wore her expensive fur coat to a pot-luck dinner. While driving her there, her husband swerved going around a sharp turn, causing her to spill her pot of food in her lap. The insurance company paid $400 to have the fur coat cleaned and restored.

One distinct advantage of a personal articles floater is that in most cases it eliminates the need to prove the value of an item when a theft or some other kind of loss occurs. This can be particularly important with respect to such things as valuable jewelry, fine arts and antiques. Without this "scheduled" coverage, you might have trouble convincing an insurance adjuster that a wall of your burned-out $80,000 house was the repository of a Picasso with a five-figure price tag.

Newly acquired property of a class already insured under a personal articles floater is automatically covered for up to 30 days (90 days for fine arts), providing that you report the addition to your insurer within that period and pay the appropriate premium, retroactive to the date of acquisition.

The personal articles floater generally imposes one geographical limitation. Fine arts are covered only within the limits of the continental United States, Hawaii and Canada. Everything else is protected anywhere in the world.

The cost of the personal articles floater will vary from place to place, and the rates in the accompanying table are listed only for purposes of illustration. The data, based on rates in effect in April 1982 were provided by an insurance broker in an Eastern state.

Item	Value	Rate per $100	Premium
Cameras:			
Canon AE-1 Camera	$190		
Autowinder	85		
200mm, F4.0 lens	140		
35-105mm zoom lens	225		
Miscellaneous filters and accessories	95		
Totals for Camera	$735	$1.65	$12.13
Golf Equipment:			
Matched sets of clubs and bags (two)	$800	$1.10	$ 8.80
Total, all items			$20.93

Although floaters are relatively inexpensive, and the total outlay is surprisingly small, you can still save on your coverage if some of your things are stored in a bank safe deposit box. Most companies will give you a substantial discount on your floater premiums for those items that are kept in your deposit box.

CONSUMER TIPS

Policy Provisions

- ☐ Recognize that your homeowners policy places strict limits on what your insurance company will pay for certain specified items.
- ☐ Talk to your insurance representative about special floaters for items such as valuable jewelry, antiques, and stamp collections.
- ☐ Secure an appraisal certificate for items for which you have no sales receipts. Ask your insurance advisor to recommend a reputable appraiser, or locate a professional by checking the Yellow Pages for persons listed as members of the American Society of Appraisers.
- ☐ Arrange for extra coverage for your valuables in either of two ways:
    - ☐ increase the policy limits on the various categories of items.
    - ☐ purchase a personal articles floater as tailored coverage to insure each article.
- ☐ Remember, it is much easier to prove the value of an item when it's in your possession than after it's lost through theft or fire.
- ☐ Ask about a separate policy for valuables you keep in a safe deposit box.

CHAPTER 6

DISASTERS MOTHER NEVER TOLD YOU ABOUT

When you were tiny your mother warned you about such perils as fire, leaving the doors unlocked for burglars, talking to strangers, and a host of other things. But Mother probably never mentioned several very special perils that can take away a goodly portion of your assets if you're not careful. Floods, earthquakes, owning a house in a neighborhood where the crime rate is high, and even the chance of losing title to your house, are all perils no one thinks about but are quite real and can be surprisingly damaging.

Losing the Title to Your House

The title to your home may be recorded in the official records at your county courthouse, but that doesn't necessarily mean it's secure. You may not be aware of it, but you actually can lose everything you have invested in your house without ever being hit by a tornado or some other natural disaster. The reason is the deeds for real estate parcels sometimes are not as clear-cut as they should be. So, when you buy a home your mortgage lender probably will require that you buy not just homeowners coverage but title insurance as well.

The purpose of title insurance is to protect you against a financial loss if someone discovers later that your title, or documentation of ownership to the property, is somehow defective. Sometimes these defects do turn up. Sometimes they go years back, and they involve all sorts of situations, such as when a previous owner failed to satisfy a judgment or a mortgage on the property, or if the house was once sold under a forged deed or an invalid will.

It's probably unlikely that such a thing will happen, but a title flaw can be buried in any one of a multitude of records which by law must be officially recorded—deeds, mortgages, judgments, ancient court proceedings, taxes, assessments and other legal instruments.

Suppose you bought title to a house from someone who owed a large amount of money to a creditor who had obtained a judgment and filed an attachment against the property. Although you were unaware of the judgment, you could lose the investment if the creditor then exercised his right to take over the property in satisfaction of the debt.

Or take the case of five Long Island, N.Y., families who built homes on land which was mistakenly included in a larger land parcel sold to a developer. The sale was a mistake because the same land had been deeded earlier to an heir of the original owner, a fact which had gone unnoticed by the developer. A court ruled that the five families had to return the land, with all the improvements they had made, to its legal owner, the heir. Fortunately, the losses were covered by title insurance. Fortunately again, the holder of the deed accepted a hefty financial settlement from the insurance company, making it unnecessary for the owners to give up the homes they had built.

In other cases, a title insurance company paid for:

- A $75,000 loss on a policy insuring 15,000 acres of Tennessee mountain land which did not exist.
- A Texas loss that came about because someone moved a road, cutting 42 acres off from the insured premises.
- Legal costs of defending a policyholder against the claims of a neighbor who asserted water rights to a well on his property.

- A $250,000 loss to a corporation whose ownership of an apartment complex in Georgia was found to be impaired by a previous mortgage on the property. The corporation hadn't known about the earlier mortgage because a county clerk had neglected to index it in the real property records.

In many states, the law requires every title insurer to maintain a complete archive of records needed to establish the credibility of each title. Records are collected from local government departments, courts and the federal government, and from any other source that could affect the title. Many of them go back 200 years or more.

Using these archives, insurers research all aspects of a title before they issue a policy. That means your chance of buying a defective title are relatively slim. It also means any defects the insurer discovers that increase the chance of loss will normally be spelled out in the policy and excluded from the coverage. Once the policy is issued, the title company assumes all losses from any loopholes and flaws in the title that it failed to find. Of course, the insurer's liability is limited to the amount specified in the policy.

The policy covers only defects which exist at the time it is issued, not those that might occur later. But it remains in effect as long as you or your heirs own the property. And in addition to covering any losses due to title defects, the insurer agrees to defend you against any lawsuit involving the title.

Title coverage makes the property more readily salable. Moreover, it facilitates getting a mortgage to finance the purchase. Title insurance frees both the borrower and the seller, as well as the lender of the mortgage money, from worries about possible losses resulting from defects or faulty title examinations.

The premium normally is a modest one-time fee, and it is usually paid at the time you purchase the house. Modest or not, the policy could be as valuable to you as the deed to your home. So, whatever you do, don't forget it or mislay it. Keep it right with the deed itself in your safe deposit box.

Being Wiped Out by Flood or Earthquake

Floods and earthquakes can wipe out your home, and since neither is covered under most policies, special insurance is available to cover either of those perils. The federal government provides flood insurance, and private insurers offer earthquake coverage.

Earthquake insurance usually is written as an addition to homeowners coverage or some other property insurance policy. Normally, separate deductibles are applied to losses involving your home, other buildings on your property and your personal belongings. If you decide to pass up this coverage, you're not alone. Although most states are vulnerable to earthquakes, they strike most often in the Pacific Coast area and more than half of all policies are purchased in California.

Because flood insurance is marketable only to people who live in areas likely to be flooded, commercial insurers have found it is generally uneconomical for them to provide that coverage. Flood insurance is the most misunderstood of all property coverages. Almost invariably after a flood there are protests from property owners who suddenly learn they are not insured and their insurance companies are not going to pay for the damage. That's why it pays to know what you're covered for in advance.

Although most policies issued by private insurers do not pay for damage to fixed property caused by flood—whether from a tidal wave or the overflow of a nearby creek—you'll probably find that flood insurance is available—from Uncle Sam. This insurance covers not only flood damage, but closely related perils, such as a mudslide set off by excessive underground moisture.

Flood insurance provided by the Federal Insurance Administration, a department of the Federal Emergency Management Agency (FEMA), is available to owners of property in flood-prone areas of every state, the District of Columbia, Puerto Rico and the Virgin Islands. It is offered in communities which qualify by setting up flood plain management—or land-use—programs that protect new construction from future flooding.

Of some 20,000 communities around the country which are considered flood-prone, more than 17,000 had qualified as of mid-1982. You can find out whether your community qualifies by asking your insurance agent.

A community can qualify for an emergency insurance program if it adopts a preliminary flood plain management program. When it goes further and adopts a more comprehensive plan, it will qualify for a permanent insurance program.

If your community qualifies for the emergency program, you may buy up to $35,000 of coverage on a single-family dwelling. If you own some other kind of residential structure you may purchase coverage of up to $100,000. The contents of your home may be insured separately for up to $10,000. For the owner of a one-family dwelling, the cost of the coverage in the emergency phase, anywhere in the country, is 40 cents per $100 of insurance.

In the permanent program, single-family homes may be insured up to $185,000 and other residential structures up to $250,000. Coverage on contents is limited to $60,000. This advanced program has a wide range of premiums which are geared to the type of construction and the location of the property.

Whenever you have a flood claim, your payment will be reduced by a $500 deductible on each building and $500 on its contents. Not only that, but the government will pay only $250 altogether for your jewelry, art objects or any precious metals you might have on hand. If a flood threatens, get those possessions away first.

Residents of communities which fail to participate even though they have been designated as flood-prone are ineligible for the federal flood disaster assistance which sometimes is authorized in the aftermath of a severe flood. As a property owner, you stand to lose out on that kind of assistance if federal flood insurance is available to you and you fail to buy it. You should organize your neighbors and urge that your community adopt a land-use program.

Many Connecticut home owners and businesspersons found themselves unprotected in the spring of 1982, when floods caused more than $100 million in damage to business properties alone, and comparable damage to thousands of homes. Federal flood insurance was available in the area, and some home owners had bought it. But, according to a National Flood Insurance Program administrator, most business firms hadn't bought the insurance "because they thought they didn't need it."

With that lesson in mind, be careful where you buy your vacation home. Building of condominiums and cottages on barrier beaches from Maine to South Carolina is said to be tapering off because the government considers those areas too flood-prone to qualify for federal insurance.

One other warning: Homes built over water or below a high tide line are not eligible for flood insurance. And there is no coverage on unanchored mobile homes in special hazard areas or for mobile homes placed in parks that were opened in coastal high hazard areas after April 1, 1982.

Although flood insurance policies are provided by the government, you can get one through any licensed property insurance agent or broker.

Losses in a High-Crime Area
Residents of large cities have turned out to be the principal beneficiaries of a Federal Crime Insurance Program which came into being in 1971. Congress set up the program to make burglary and robbery insurance available to people who are refused such coverage in the private market because their neighborhoods are plagued by severe crime conditions. The policies are sold only where the federal insurance administrator determines that it's difficult to get insurance.

Although 27 states, the District of Columbia, Puerto Rico and the Virgin Islands were participating in the program in 1982, New York alone accounted for more than 60 percent of all the policies and almost 70 percent of the dollar amount of insurance in force. For the most part, the states which have the program are the same as those which operate programs called FAIR Plans.

Under the Federal Crime Insurance Program, you may purchase up to $10,000 of coverage against financial losses resulting from robbery or burglary, subject to a deductible of $100 or 5 percent of the gross amount of the loss, whichever is greater. However, the insurance will pay no more than $500 each time there is a theft for losses of such valuables as jewelry, articles of gold, silver or platinum, furs, fine arts, antiques, and stamp or coin collections.

To qualify, home owners or renters must have door and window locks that meet government specifications. The annual premium for household crime insurance ranges from $60 for $1,000 of coverage to $120 for $10,000 of coverage. The insurance, where it is available, is sold by licensed property insurance agents and brokers.

When Your Property Is "Uninsurable"

Insurance companies reserve—and sometimes exercise—the right to refuse coverage to owners of homes and commercial properties where there is an unusually high risk. To accommodate the property insurance needs of people who have difficulty obtaining coverage, the industry operates "Fair Access to Insurance Requirements" plans (familiarly known as FAIR Plans) in 26 states, the District of Columbia and Puerto Rico. They are supported by insurance companies writing insurance voluntarily on other properties in those states.

These plans were established by Congress in the wake of the widespread riots and civil disorders in the 1960s. There had been staggering amounts of property damage, particularly in the inner-city areas of major cities, and many insurers were forced to tighten their underwriting criteria, making it hard to get coverage in some areas.

The FAIR Plans are offered mainly in blighted or distressed areas, as an insurance industry-supported effort to make insurance available. Owners of properties in such areas often have difficulty getting insurance not because of the location but because they often fail to keep their properties maintained and they decline to eliminate hazards that literally invite losses. Industry leaders point out that where local associations get together to improve the neighborhood, as in Brooklyn's Bedford-Stuyvesant area, insurance has become more readily available.

Even the FAIR Plans aren't required to accept all applicants. They have the option of rejecting an application, accepting it outright or accepting it on condition certain improvements are made. Inspections for insurance purposes are made at no cost to the applicant.

For example, a home may be rejected if it doesn't meet minimum fire safety standards, or if it is in violation of building or safety codes. If you want coverage on a vacant building, you'll probably be required to board it up first or take other security measures.

While faulty wiring that threatens to cause a fire almost any minute would be unacceptable, a house with a junk pile full of combustibles might be accepted if the owner agrees to clean it up. If a property doesn't meet the criteria of a FAIR Plan, a representative of the plan will advise the owner about what he or she can do to make the place insurable and will provide a list of companies where the owner can get coverage.

The FAIR Plans are not permitted to turn down an application for insurance strictly because of a property's location or because of environmental hazards over which the property owner has no control. If an applicant for coverage isn't satisfied with a plan's underwriting decision, he or she may appeal to the plan's governing committee. Additionally, disgruntled applicants have recourse to their state's insurance department and, if necessary, to the courts.

Actually, the FAIR Plans reject fewer than 1 percent of owner-occupied houses, according to the Property Insurance Plans Service Office, which provides advisory services to the plans. The rejection rate is a little higher among dwellings which aren't occupied by their owners.

Usually, people who have trouble obtaining coverage in the standard market in FAIR Plan states or elsewhere, or who want more insurance than a FAIR Plan will provide, can get the coverage they want from a company which specializes in high risks.

In terms of price, buying insurance through a FAIR Plan is similar to doing your grocery shopping at a "mom-and pop" store rather than a supermarket. The price usually isn't cut-rate, but it's generally in line with the premiums charged by many individual insurers. High-risk insurers, for obvious reasons, usually must charge more than the standard rates for property insurance. The owner of a home which is accepted conditionally by a plan probably will be charged a higher rate, at least until the necessary repairs are completed or the dangerous condition is corrected.

You can apply for insurance through a FAIR Plan with any licensed insurance agent or broker, or you can contact an office of the plan.

Riding Out the Hurricanes

Similar to the FAIR Plans are beach and windstorm plans operated by property insurers in seven states along the Atlantic and Gulf Coasts. They provide insurance against damage from hurricanes and other windstorms. Those plans operate in Alabama, Florida, Louisiana, Mississippi, North Carolina, South Carolina and Texas. As certain coastal areas are notoriously subject to hurricanes and other ocean storms, insurance on properties in these areas often had been hard to get. The beach and windstorm plans, like the FAIR Plans, were set up to deal with the problem. The insurance is similarly available through any licensed agent or broker, or through plan offices in the seven states.

Generally speaking, if you can't get property insurance either from the FAIR Plans or the "high-risk" market, or flood insurance from the federal government, your house probably is "uninsurable." In other words, the probability of a major loss is so high no insurer is willing to accept it, because it would not be feasible to charge a high enough premium to pay for the predictable losses or to place the property on a reasonably equal risk-sharing basis with other insured properties. Before you agree to buy a property that could fall into this category, check. Don't put your money into an asset that's uninsurable.

CONSUMER TIPS

Essential Protection

There are some points **not** covered by your homeowners policy. Not all of them are for everyone, but some just may be crucial in protecting what's yours. At any rate, it's smart to make a conscious decision about your need for each kind of coverage.

What's **not** smart is to assume you're covered only to find out you aren't after you've had a loss. Keep these important points in mind and discuss your concerns with your insurance representative.

TITLE INSURANCE
- ☐ Protects you against financial loss if your documentation of ownership of your home or property is at issue.
- ☐ Could become as valuable as the deed to your home. Keep your title insurance policy with your deed. And keep them both in a safe deposit box.

FLOOD INSURANCE
- ☐ Is made available by the Federal Insurance Administration to owners of property in flood-prone areas.
- ☐ Can be obtained through any licensed property insurance agent or broker.
- ☐ Is a key part of eligibility for federal flood disaster assistance which sometimes is authorized by the President in the aftermath of a severe flood. As a property owner, you stand to lose out on that kind of assistance if federal flood insurance is available to you and you fail to buy it.

EARTHQUAKE INSURANCE
- ☐ Is provided by insurance companies as a special addition to your homeowners or other property insurance policy.

CRIME INSURANCE
- ☐ Congress set up the Federal Crime Insurance Program to make burglary and robbery insurance available to people who were refused such coverage in the private market because of severe crime conditions that exist in some areas.
- ☐ Policies are sold only in locations where the federal insurance administrator determines there is an insurance availability problem.
- ☐ Crime insurance is sold in 27 states, the District of Columbia, Puerto Rico and the Virgin Islands by licensed property insurance agents.
- ☐ FAIR Plans have been organized as an insurance industry-supported effort to make insurance available to property owners.

CHAPTER 7

INSURANCE FOR THE BARNYARD

If your "home, sweet home" is a "home on the range"—a farm or ranch—you'll need insurance coverage for a lot more than what's under your own roof. There are a lot of other things out there you need to protect—things like barns, silos, farm vehicles, equipment, and even crops in the fields. These are kinds of property that the standard homeowners policy isn't cut out for.

Liability questions can be tricky, too. Suppose a visitor falls over a tractor attachment. Is the farmer personally liable? Or is it his business that bears the liability?

To resolve such problems, insurance companies have developed farmowners-ranchowners policies that combine personal and business coverages in a single package. These policies are similar in design and content to those used for homeowners except that they come in several parts—one for the farmhouse and personal property and another to cover the farmer's business assets.

For his home and personal property, the farmer can tailor his coverage to his needs by choosing from three types of coverage: basic, broad and tenant's.

Basic Coverage—This covers the farmhouse and all personal property, as well as providing for additional living expenses or loss of rental income should the house be damaged. The policy covers losses by fire, theft, windstorm, hail, explosion, riot, civil commotion, aircraft, vehicles, smoke, vandalism and malicious mischief as well as damage to personal property if a vehicle turns over. It also covers loss if property has to be removed because it is endangered by one of the perils for which the farmer is insured.

While this coverage is similar to that offered under the basic homeowners policy, there are some differences. Most outbuildings are not covered. The only structures which are covered are private garages (for 10 percent of the amount of coverage on the house). Lawns, shrubs, trees and broken windows are not covered.

Broad Coverage—This insures the farmhouse and personal property against the same perils listed in the broad form of the homeowners policy. (*See Appendix B*).

Tenant's Broad Coverage—If you rent a farm or a farmhouse, you can use this policy to cover your personal property against the same perils as the broad coverage policy. It doesn't cover the dwelling, since any losses would be borne by the owner.

Farm Equipment and Livestock

In most areas, farmers and ranchers using the package policy have two options for insuring their "farm personal" property, including equipment and livestock but excluding farm buildings. One option, which is available in all states, permits the farmer to itemize—or "schedule"—such property. The second choice, available in some areas, is a blanket (non-itemized) form of coverage.

Both insure against fire, lightning, removal of debris, windstorm, hail, explosion, riot or civil commotion, aircraft or vehicle collision, smoke, theft, vandalism and malicious mischief. They even will reimburse the farmer for livestock electrocuted by electrically charged fences, broken power lines or other voltage sources "dumb animals" couldn't be aware of.

You may use the first, or itemized form of farm personal property coverage to insure most property usual to the operation of a farm, including hay, grain, fertilizers, machinery, farm vehicles and equipment, farm records and livestock. In some cases, the coverage may be extended to stored farm products.

The second form gives blanket coverage to most of the farmer's personal property. Most, but not all. While the blanket coverage applies to goats, horses, mules and donkeys, other animals, such as dairy cows, beef cattle and poultry are not included. Nor are certain crops, such as tobacco, cotton, vegetables and fruit. The blanket form generally is the more expensive of the two types of coverage.

Farm Buildings

Barns and other farm structures, including dwellings other than the main farmhouse, are covered separately, under another section of the farmowners-ranchowners package.

Liability

The farmer can add liability coverage to his farmowners-ranchowners policy that is similar to the protection offered in the homeowners policies, except that it also covers claims stemming from his business operations. This insurance would cover a claim by a person who was injured on the farm, whether the accident happened in the farmhouse, in a barn or in the open.

If you're a farmer and you operate a roadside stand, your liability insurer will even stand behind your products, so long as they are agricultural. For example, the coverage would protect you against a claim by someone who became ill after consuming a tainted fruit or vegetable that had been purchased at your stand.

Under the same liability coverage, the insurance company will pay medical costs of household workers whose duties do not relate to the farmer's business, if they are injured on the property. In some states, this coverage may be extended to farm workers as well. For an extra premium, the farmer can have the medical payments provision expanded to include family members who might be injured while working on the farm. But if you already have a good family medical plan, you shouldn't buy this option.

Optional Coverages

Some other options are worth considering as well. One is for animal collision. If a cow, horse, hog, sheep or goat is hit by someone else's vehicle on a public highway, the insurance company will reimburse the farmer up to $400 per animal.

Two of the most popular of the other options a farmer may add are all-risk coverages for items in which farmers frequently have large investments—mobile agricultural equipment and livestock.

These endorsements require a listing of all the property to be covered. The property then is insured against every peril except those specifically excluded under the terms of the policy. So there are no misunderstandings later, you should know the exclusions, which include loss by mysterious disappearance and accidental shooting by farm workers or hunters. If an animal escapes, it's not covered. And losses not covered by the equipment endorsement include damage by wear and tear, severe temperature changes, mechanical or electrical breakdown, or damage resulting from lack of maintenance.

Farm Autos

If you are engaged in farming or ranching, you may qualify for a discount on your auto insurance premiums for a private passenger car, station wagon, a motor vehicle with a pickup body or a panel truck. To be eligible the vehicle may not be used in any occupation other than farming or ranching, nor for driving to and from any other occupation. If you or your children commute in a farm vehicle that's insured under such a discount, beware. You might not be covered should you have an accident.

Crop-Hail Insurance

Until the time comes when weather patterns can be controlled or modified, insurance offers just about the only protection a farmer has against financial ruin because of damage to crops by the elements. Crop insurance, unlike most coverages, is available both from private insurance companies and the Federal Crop Insurance Corporation, an agency of the U.S. Department of Agriculture.

A variety of crop-hail policies is available. Coverages, which generally differ from state to state, are tailored to allow for differences in climate and other environmental factors. In a crop policy, coverages are limited to acreage rather than the total market value of the crop. Usually, the coverage becomes effective once the crop is above the ground and expires when the crop is harvested.

The policies generally are available on an all-risk basis. In other words, crops are insured not only against damage by fire and severe storms, but against such additional perils as insects, excessive moisture, drought and any other peril not specifically excluded.

An illustration of the unpredictability of hail losses was a hailstorm that hit a Wyoming farm area in early summer. Just as adjusters were writing checks to settle the claims another big hailstorm hit, less than two weeks later. Adjusters coped with the situation by estimating the new damage in total, subtracting the amounts already paid, and writing the farmers additional checks for the difference.

Such protection is a far cry from the Old West days when every catastrophe was an "act of God" and there was little any farmer or rancher could do about it but take off his ten-gallon hat, get down on his knees, and pray.

CONSUMER TIPS

Barnyard Basics

If you own a ranch or a farm, take a look now at your insurance policy.
- ☐ Check to see exactly what is—and is not—covered.
- ☐ Find out whether liability coverage is adequate and up to date.

Which form of the policy do you have?
- ☐ Basic Coverage
- ☐ Tenant's Broad Coverage
- ☐ Broad Coverage

Does your policy cover your needs?

Check the coverage on your livestock and equipment to see how you have elected to insure your "farm personal" property:
- ☐ Property is scheduled, itemized.
- ☐ Property is covered in a non-itemized, blanket form.

Do you have:
- ☐ Livestock or equipment which might require modifying your insurance coverage?
- ☐ Proper coverage on barns and other structures?
- ☐ Up to date coverage on outbuildings?

Check the liability portion of your policy.
- ☐ Do you need to consider raising or lowering the amounts of your coverage?
- ☐ Are you taking advantage of discounts for which you may be eligible?
- ☐ Have you made a conscious decision about your need, or lack of need, for crop-hail insurance?

CHAPTER 8

WHAT TO DO WHEN YOU HAVE A CLAIM

A Florida man called his insurance agent recently and announced that he was putting in a claim for a loss. But the agent replied that the loss, as described, was not covered. The policyholder quickly changed his story, offering a different version of the incident. No, said the agent, he was sorry but that was not covered either. Finally, in desperation, the caller said, "Well, you tell me what is covered, and I'll tell you what happened."

The man might have been frustrated because he had not insured himself to cover that particular loss, but that's definitely not the way to file a claim with your insurance company. If you do, you can wind up in trouble. For that reason, and simply to make it easier for you to get the money you are due, it's important for you to know what to do when a loss does occur.

First, call your insurance agent or your insurance company. Most home insurance policies require you to notify your insurer or your agent immediately if you have a property loss that's covered. But regardless of what your policy requires, call your agent anyway. He can help. In fact, that's one of the services you pay for when you send in your premium each year. For one thing, he'll tell you right off whether you're covered. If not, he'll explain why. And if so, he can help you get your claim processed promptly. The peril that caused your loss might not be among those for which you are covered, or the amount of damage might not be as great as your deductible. He can tell you quickly and relieve you of any uncertainty. On the brighter side, he might find you have coverage you didn't even know about.

For instance, not long ago a Wisconsin woman was browsing in a Mexican antique shop and knocked over a Chinese vase. Unfortunately the vase was worth $11,500. When the shop owner presented the woman with a bill for the loss, she filed a claim with her insurance company back home. Normally one would not expect their homeowners policy to cover them in far-off Mexico, but in this case the woman's homeowners liability insurance paid the entire $11,500.

If your loss is covered, or if you're not satisfied that it's not, follow up your call with a written notice telling your insurance company what happened and the nature of the damage. Later, the company might ask you to submit a sworn statement with such particulars as an inventory of those personal articles that were lost and estimates for repairs.

Always report any burglary or theft to the local police as well as to your insurance company, and report any loss involving a credit card to the firm that issued the card. If you fail to do either it could invalidate your coverage.

One policyholder found this reporting detail a bit of a problem, but he solved it ingeniously. Sometime during a plane flight from Tulsa, Okla., to Wichita, Kan., his expensive overcoat was stolen. To complete his insurance claim, he had to notify police—but he didn't know exactly when or where his coat was taken. His solution was to take a map and trace the flight by drawing a line between Tulsa and Wichita. Finding a town that lay on that line, he notified the sheriff there by mail and got the theft officially reported. The insurance company paid the claim.

As a practical matter, once you have reported a loss, your insurance representative probably will assist or advise you. Among other tasks that fall to you:

- If your property is damaged, whether by fire or some other insured cause, do whatever has to be done to protect it from further damage. Should a burglar break a lock on an outside door, repair the lock rather than take a chance of leaving the house unprotected. Board up broken windows. If a fire burns a hole in the roof, make whatever temporary and reasonable repairs are necessary. And don't forget to submit to your insurer for reimbursement the receipts for what you spend.
- Obtain estimates covering repairs to structural damage.
- Prepare an inventory of lost or damaged personal articles. Include a description of each item, along with its present value and what you figure to be the dollar amount of the loss. Attach bills, receipts and other documents that substantiate your figures.
- Be sure to keep a careful record, including receipts, of any additional living expenses you incur if you have to find other accommodations while your house is being repaired. If you have to move out, it's advisable to let your agent or other company representative know your plans in advance.

Once your company has determined your loss is covered, it probably will assign an adjuster to verify your claim and determine the amount of the loss. Most claims are settled promptly, but some require prolonged investigation, often because of the extent of the loss or because its cause is unclear. If you want an independent evaluation, public adjusters are available, for a fee.

Insurance companies recognize that their adjusters and their policyholders don't always see eye to eye on the proper sum for a settlement, and you don't have to accept an amount you don't think is fair. If you and the company can't reach agreement, either of you can demand that the dispute be submitted to arbitration. Each then designates and pays for the services of an arbitrator and the two arbitrators select a third disinterested arbitrator, whose fee is split between you and your insurer. Whatever amount any two of the arbitrators agree on is binding.

Often, the most difficult claims to handle are those involving extensive loss of personal belongings. You should always keep an up-to-date household inventory, and attach the necessary receipts and other documents supporting what you say your possessions are worth. If you've done that, your problem is well on its way to resolution.

If you have to call on your memory, you have the makings of a real headache. You'll probably overlook some items and you may have trouble coming up with realistic valuations of lost articles. The adjuster probably will accept your statement that a dining room table and matching chairs were among the items that perished, but if you claim to have bought them at a high-priced furniture store you'd better be able to prove it if everything else in the house clearly came from discount marts. If you claim an item is a valuable antique, you'd better be able to trace it to a dealer, an auction house, or your great-grandmother, rather than to Joe's Second-Hand Furniture Emporium.

This is when a written inventory, drawn up in advance, is invaluable. People always forget things after a loss. One agent reports that when a client gives him a list of personal items after a fire, he sends the client back to make a new list. The policyholder always comes back a second time, often a third or fourth time, with a longer list. The agent believes it's his duty to put policyholders through this routine, because they are entitled to recover everything that's due them.

While some policies today provide for the replacement of lost or damaged articles with new items of similar kind and quality, policies more commonly pay off on the basis of actual cash value. Actual cash value is generally the equivalent of an item's replacement cost, minus depreciation.

Assume that fire destroyed a sofa which you purchased five years ago for $400 and which was in generally good condition at the time of the fire. Say that the same piece of furniture, or one of like kind and quality, would cost $600 today. Under the replacement cost coverage, you could expect to recover the price of the new sofa. (If the sofa was damaged but not destroyed, your insurer might reimburse you for the cost of repairing it rather than replacing it.)

Under the actual cash value method, a depreciation factor would be applied to that $600 replacement cost figure. If your sofa normally would be expected to hold up for 10 years, it would have depreciated 50 percent in the five years you had it and you would be allowed about $300 (or 50 percent of $600). If the life expectancy of such a sofa were 15 years instead of 10, it would have depreciated by one-third rather than by one-half, and you could hope to collect $400.

Whatever settlement procedure your policy provides for your personal property usually applies also to carpeting, domestic appliances, awnings, outdoor antennas and outdoor equipment regardless of whether they are attached to buildings.

As explained earlier, partial damage to your house or to another building on your property is fully covered—up to the limits of the policy—if you have insured the house for at least 80 percent of its replacement value. If you have less coverage, your reimbursement will be figured downward in proportion.

Usually, if the damage is extensive, insurers won't pay the full replacement cost until the repairs are completed. Until then, you are entitled only to the actual cash value for the damage, with the balance to come after the work is finished.

You may conclude you don't want to repair or replace the damage at all, or you may decide to put off such a decision until later. In that case, you can put in a claim for the actual cash value of the damage and accept a settlement on that basis. If you decide to go ahead with the repairs, you have 180 days after the loss to put in a claim for the difference between the actual cash value and the replacement cost.

Liability Situations

Picture this common type of occurrence. You have a few neighbors in for a party. Your porch light had burned out the day before. As he's leaving, one of your guests slips on the bottom step—perhaps on a hidden piece of ice—and falls flat on his face on the sidewalk. He says he's okay, but his face is scratched up and you notice a welt on his forehead.

Never mind that your neighbor said he was feeling fine. For your own protection, get on the record by reporting the incident to your insurance agent or company in the morning. Your neighbor might turn out to have a concussion, or he might start having dizzy spells three months later, attribute them to the fall, and file a claim or a lawsuit against you as the responsible party.

Follow up your initial call with a letter explaining what happened, when and where it happened and who was hurt. Policies covering your liability to others usually require written notification as soon as practicable after an accident. Be sure to include the names and addresses of any witnesses to the accident and be completely truthful in regard to the facts. If you were at fault in not replacing the light bulb, give your insurer that information. If your neighbor was clowning around on the steps, say so.

In the event of an injury covered under your policy, your insurer will pay the expenses of any first-aid that may be required, but be careful about making any further commitments to the injured person. Your insurer is not liable for any settlement agreement or other arrangement you make independently. For example, if the neighbor who slipped off the porch step discovered he was more seriously injured than he thought and you agreed to pay his medical bills, you might find yourself footing those bills out of your own pocket. You're insured—have him file a claim with your insurer.

By the same token, if you were to offer a reward for the return of a valuable stolen painting, your insurer would not be liable for payment of the reward.

When a claim or a lawsuit is filed against you, forward all notices, papers and other documents to your insurance company as soon as you receive them. Don't forget to make copies for yourself first. And remember, your insurer will go to bat for you and will expect from you whatever cooperation it needs in that endeavor. That applies, of course, whether the claim stems from someone else's injury, or damage to someone's property, for which you're alleged to be responsible.

Medical Payments to Others

A claim under the medical payments section of your homeowners policy requires written proof of loss to the insurance company by or on behalf of the injured person. This coverage, you'll recall, will pay for medical and related expenses of injured persons regardless of fault. As with other kinds of losses, start with a phone call to your insurance representative.

The insurance company probably will ask the injured person for authorization to obtain copies of medical reports and records. And it may require a physical examination by a physician of its choosing.

Damage to Property of Others

If your 3-year-old manages to splash paint liberally over a friends's living room wall, or if you accidentally set fire to a visitor's new alligator purse, this part of your policy is available to pay for the damage. Your agent or company will ask you for a sworn statement—you have 60 days to submit it, although you'll probably want to take care of it right away—and may ask for an opportunity to examine the damaged property.

There are a couple of points to remember in making a claim:

One is that insurance adjusters generally try to be fair. If you don't think one is, he usually can be reasoned with. One case in which an adjuster admitted he was not an expert was that of a woman whose expensive lingerie was stolen from her clothesline. She gave the adjuster a complete list of the stolen items, for which the adjuster offered her an amount consisting of costs less "depreciation." The woman hit the ceiling, claiming the adjuster didn't know what he was talking about, that lingerie couldn't be depreciated. Taken aback, the adjuster consulted with the women who worked in his office, all of whom supported the claimant. The adjuster gave in, saying, "I'm only a man—I don't know about these things."

You should remember there is a time limit on resolution of claims, as well as a requirement for prompt reporting of losses. One claimant, who was a lawyer, had a dispute with the insurance company over additional living expenses. The company offered him $2,500, but the attorney claimed $3,500. After a year had gone by, the claimant received a letter from the insurance company saying the statute of limitations had run out. The lawyer should have known better, but because he neglected his own legal affairs, he failed to collect either the $3,500 or the $2,500.

In a similar case, the same company paid the claim because the case was in arbitration at the time the statute of limitations ran out. In that situation the claimant was still actively doing something about his claim.

Another point to remember is that the settlement you receive or are offered may have little to do with what a friend or neighbor may have claimed to have received in a similar situation. An Ohio agent reports that after a tornado ripped through the town of Xenia a few years ago, his office was filled with as many as 50 people at a time, all neighbors, all waiting for settlements on their destroyed homes. But the settlement offers varied, even for houses of comparable value. Some claimants got upset about this, until the agent patiently explained they were insured under a variety of policies with different provisions, including several different basic forms of the homeowners policy, each of which provides a different range of coverages and exclusions. The agent's advice to policyholders is to understand your own policy before you have a claim. And that's good sense. It could save you a lot of money when you have to make a claim.

CONSUMER TIPS

What to Do When You Have a Claim

Follow these steps for best results in getting maximum value for your home insurance dollars—an important part of protecting what's yours.

- ☐ Phone your agent or company.
    - ☐ Are you covered?
    - ☐ Does your claim exceed your deductible?
- ☐ Follow up your call with an explanation of what happened in writing.
- ☐ Report any burglary or theft to police.
- ☐ Ask questions and get any needed advice from your insurance representative.
- ☐ Make temporary repairs and take any other steps necessary to protect your property from further damage. Save receipts for what you spend and submit them to your insurance company for reimbursement.
- ☐ Obtain estimates for repairs to structural damage.
- ☐ Upon request, submit a statement including an inventory of lost articles, repair estimates and a record, including receipts, of any additional living expenses you incur if you have to find other accommodations while your home is being repaired.
- ☐ Provide needed information to the adjuster assigned to handle your claim.
- ☐ If you disagree on the amount of the settlement offered by your company, you may wish to submit the dispute to arbitration.

What to do When You Have a Claim

What to Do Before a Claim

- ☐ Make a household inventory.
- ☐ Keep the inventory up to date and attach receipts, appraisal certificates and any other documents supporting the value of your possessions.
- ☐ Keep a copy of your inventory in a safe place **away** from your home.
- ☐ Look closely at your home insurance policy. Are your belongings insured for:
    - ☐ actual cash value (replacement cost of an item minus depreciation)?

    – or –

    - ☐ replacement cost (the amount it would take to replace the item at current prices)?

Before you have a claim, it's in your best interest to find out exactly what your policy provides. (That way, **you** can change anything which does not suit **your** needs or expectations **now** and thus avoid disappointment in the event **you** have a loss.) After all, protecting what's **yours** is top priority here!

- ☐ Is your home insured for at least 80 percent of its **replacement value**? (if you have less coverage, you won't be able to collect full reimbursement for any partial damage.)
- ☐ Do you understand the liability coverages provided in your homeowners policy? (If you have questions, now's the time to ask your insurance representative for answers.)
- ☐ Remember, when someone is hurt on your property—or even when you **think** someone **might** have been hurt—follow these steps:
    1. Report the incident to your insurance agent or company by phone as soon as possible.
    2. Follow up your call with a written account of what happened.
    3. Be careful about making any settlement commitment on your own. Instead, have the injured person file a claim with your insurer.
    4. Forward all notices, court papers and other documents to your insurance company as soon as you receive them. Keep copies of everything for yourself.

How about the medical payments portion of your homeowners coverage?

- ☐ This type of claim requires written proof of loss.
- ☐ It pays for medical expenses of an injured person regardless of fault.
- ☐ A medical payment claim begins, as do others, with a call to your insurance representative.

What if you are responsible for damage to someone else's property?

- ☐ Phone your agent.
- ☐ Submit a sworn statement of the loss within 60 days.
- ☐ Allow the adjuster to inspect the damage.

Here are three general points to remember about home insurance claims:

- ☐ Adjusters try to be fair, but sometimes do make mistakes. Talk things over with your agent and adjuster if you are dissatisfied with the settlement offer.
- ☐ Insurance policies place a time limit on resolution of claims. Be sure to keep your claim active or settle it before the resolution period expires.
- ☐ The coverages and exclusions in your insurance policy can differ significantly from those of your friends. The best advice is to understand **your** policy before **you** have a claim.

CHAPTER 9

HOW TO AVOID DISASTER

If you have a loss covered by an insurance policy, the insurance company will be there. Assuming you knew what you were doing when you bought your insurance, you'll be protected against the worst financial consequences. All that notwithstanding, the best insurance of all is prevention.

Where you can't avoid losses, you often can limit them by taking some basic precautions beforehand. A little prevention can save you not only the inconvenience of the loss itself, but even injury or death.

If you practice them generally, safety and security also are good ways to keep insurance premiums down. Fewer claims mean lower costs for insurance companies, so they can afford to compete for your business by charging you less for coverage.

Let's start with your home.

What to Do About Fire

If you live in your own house your biggest worry should be fire. It's by far the most destructive threat to your home. A fire breaks out somewhere in the United States about every 45 seconds, on the average. Fires claim some 7,500 lives a year in the United States, and cause property damage running into billions of dollars. Yet many, if not most, of these losses could have been prevented.

Here's what you should do.

Eliminate Potential Fire Hazards—Make sure all electrical wiring is in good shape. If not, replace it. If you need to, get a qualified electrician to inspect it and advise you. Make sure all stoves, ranges, furnaces and fireplaces are functioning properly, and that chimneys are cleaned. If you make home improvements, see that fire-resistant materials are used. Plaster and plasterboard are good fire resisters. Some kinds of wood paneling are treated for fire resistance, other kinds aren't; find out how they are rated. Use only appliances bearing the approval seal of Underwriters' Laboratories or another recognized testing laboratory. In buying a house, give some thought to ease and safety of exit in case of fire. In general, brick or masonry construction is more fire-safe than wood frame construction. Check on water supply and availability of fire services.

Other fire hazards may be created by unsafe housekeeping practices. Here are a few tips on fire-safe housekeeping:

- Keep trash in covered containers and dispose of it regularly. Don't let trash accumulate, either indoors or outdoors. Clean up attics, basements and garages regularly.
- Store paints, paint thinners and other flammable materials in their original containers—away from fire sources.
- Clean work areas of paint, sawdust or trash after every do-it-yourself project.
- Be careful not to overload electrical circuits or use frayed extension cords. Never run an extension cord under a rug or behind curtains.
- Keep home fire extinguishers at handy locations.
- Discourage children from playing with matches or candles. And if there are open-flame heaters or fireplaces in your home, make sure all nightclothes are made of non-flammable materials.

- If you smoke, keep plenty of ashtrays handy—the type lit cigarettes won't fall from—and never, never smoke in bed or while lying down on a sofa where you could fall asleep. Almost anything around a smoker can become a fire hazard.

A young Milwaukee citizen who sported a full beard found that whiskers catch fire when he attempted to consume a drink called a "Flaming 15l" at a local bar. Only the bartender's presence of mind in using a seltzer gun as a fire extinguisher saved the man from severe burns.

Install Smoke Detectors—Local ordinances and fire codes in many communities now require that houses and apartments be equipped with smoke detectors. Whether or not your town is one of them, smoke detectors are an excellent idea. They're not expensive, they're easy to install, and nearly all are highly dependable.

They have saved numerous lives. When activated, their ear-piercing wails can provide you and your family precious extra minutes to escape from a burning home, especially at night when most home fires occur. Here are some tips on their use:

- In installing smoke detectors, placement is important. At least one unit should be placed on the ceiling of the hallway leading to bedrooms.
- If your home has more than one level, consider installing a smoke detector on each level.
- If your smoke detector is battery-operated, check the batteries periodically to make sure the unit is operating properly. Typical battery operated units emit a signal when the battery is losing its power; choose one that has this feature.
- Make sure smoke detectors you buy are a type approved by Underwriters' Laboratories or other recognized testing facilities.

You have a choice of a few different kinds of smoke detectors.

The basic choice is between photoelectric and ionization types. The photoelectric type contains a bulb producing a small beam of light and a photocell which causes an alarm to sound when smoke cuts through the beam. The ionization type contains a radiation unit which produces a small electric current, and the alarm is triggered when smoke particles reduce the flow of electricity. In tests, the photoelectric type has been found to be somewhat quicker in responding to smoky fires, the ionization type to flaming fires. Consider installing one of each, a photoelectric unit near sleeping areas, an ionization unit near living areas or basement stairs.

Smoke from cooking can set off false alarms by smoke detectors. But such incidents need not cause alarm in the user, since the source of smoke is obvious. Using an exhaust fan to clear the air will stop the alarm in due course. Disconnecting the unit is not advised, as it is too easy to forget to re-connect it.

Other choices are between battery-operated units and those with a plug-in cord. Batteries obviously need periodic replacement; plug-in units would become inoperative in the event of a power failure. But both types are considered quite acceptable if these facts are kept in mind.

Smoke detectors are now widely available. You can buy them in almost any hardware or housewares store, and the only tool you'll need to install them is a screwdriver.

You can test them simply by holding up a lighted match. But don't be like one woman in a case reported from California who after successfully testing her new smoke detector in this manner, absent-mindedly tossed the lighted match into a trash container, starting a $50,000 fire.

Develop a Family Escape Plan—For your family's safety, it's best to work out in advance just what your household should do in the event of a fire. Fire safety experts recommend a somewhat formal approach:

- Sketch the layout of each floor, including windows, doors and stairways. Make sure every member of your family is familiar with the layout.
- Work out two escape routes from each room (in case one route is blocked by fire), and mark them clearly in the sketch. If there are not enough built-in exit routes, you should consider installing ropes or rope ladders at upper

windows. These should be stored in handy places, and family members rehearsed in their use.

- Hold fire drills, including some at night, so that each member of the family will know what to do and be able to act quickly in an emergency.
- Assign a member of the family to be responsible for the elderly or the very young to help them escape. Assign another person to be responsible for calling the fire department.
- Designate a meeting place outside the house and instruct everyone to go there in case of a fire.
- Once outside, count heads and then stay together. Make sure all of you have carried out your assignments. Do not go back into the dwelling for fur coats, teddy bears or other personal belongings. Let the firemen deal with the fire.

Use Care With Wood Stoves—If you're among the thousands who have succumbed to the lure of the wood burning stove, keep in mind this return to the "good old days" can have some old-fashioned drawbacks. Fire hazard is one of them. The resurgence of the wood burner has led to an alarming number of fires traceable to careless installations or misuse.

A generation gap exists between experienced wood stove users who may have grown up with them and new, first-time users unfamiliar with problems that can arise. To fill that gap, here are some wood stove do's and don'ts:

DO make sure there is enough space between the stove and combustible walls, floors and ceilings. A three-foot distance from walls is usually recommended. Check manufacturers' guidelines and local fire codes.

DO place the stove on a fireproof base.

DO have a mason or other competent professional inspect the chimney.

DO burn only dry, well-seasoned wood.

DO open a window a crack for ventilation.

DO keep the flue clean.

DO dispose of ashes in a closed metal container outside the house.

DON'T extend the stove pipe through a wall or ceiling. Have a chimney installed by a professional. Make sure it is a chimney listed by a recognized testing laboratory.

DON'T connect a wood stove to a fireplace chimney unless the fireplace has been sealed off.

DON'T connect a wood stove to a chimney serving another appliance burning other fuels.

DON'T start a stove fire with flammable fluids.

DON'T burn trash in a stove—doing so can start a chimney fire.

If you're thinking of buying a woodstove, make sure the one you select is made of sturdy, suitable material such as cast iron or steel. Beware of materials that could crack or shatter, like ceramic. Look for stoves listed by a recognized testing organization, such as Underwriters' Laboratories (UL). If you buy a used stove, check it for cracks or other defects. Also check legs, hinges, grates and draft louvers. If you live in a mobile home, be sure the stove is a type specifically approved for use in that kind of dwelling.

In installing a stove, be sure that you are familiar with local fire and building codes. Town officials should be glad to provide this information. Think twice about where you'll put your stove. Usually a central location is best. If you place the stove too near a stairwell, much of the heat will rise to the next floor. On the other hand, that may be part of your heating plan.

The National Fire Protection Association issues standards for installing and mounting wood stoves. Obtaining and following these standards would be an excellent idea.

Most of the same rules apply to coal burning stoves. But obtain the particular specifications.

Fireplaces and Heaters—A built-in masonry fireplace also can present hazards. Burn only seasoned wood, use a screen in front of the opening, and keep the chimney clean. Free-standing fireplaces require similar precautions as a woodstove; again it's a good idea to check local codes and ordinances and follow manufacturers' specifications.

Kerosene heaters are prohibited in some areas; if they are permitted in your locality, be sure you have a unit approved by a recognized testing laboratory and equipped with safety features such as an automatic shut-off switch to protect against jarring or tilting.

Whether you have a fireplace, a stove or a heater, one final word of caution. Wherever there is combustion, there should be sufficient ventilation to maintain an adequate oxygen level in the room. That's not just to keep the fire from going out—it's to keep you alive.

What to Do About Windstorms

Next to fire, the biggest threat to your home is windstorm. A hurricane or tornado has the power to demolish your house; even a glancing blow from one can do major damage. Lesser storms can topple trees or send big limbs crashing down on your house, strip shingles off your roof, or drive rain through cracks damaging floors and plaster.

Obviously you can't stop the wind, but you can make your home less vulnerable. You may think that because your home is not in the Southwestern "tornado belt," you are safe from this most common of violent windstorms. Not necessarily. Although more tornados strike Texas than any other state, no state is entirely free from the threat. A "twister" can hit almost anywhere, under certain meteorological conditions.

Before a tornado:

- Have emergency supplies on hand. First-aid materials, canned food, water and a camp stove are good to have on hand in case of a power failure.
- Keep a battery-operated radio, a flashlight and a supply of fresh batteries in a convenient place. You should have these handy in any power failure.
- Know the locations of designated shelter areas. They are most likely to be located in schools, public buildings, churches or shopping centers.
- Keep an inventory of your household furnishings and other possessions. Supplement the written inventory with a photograph of each room. This is a good idea in case of fire or burglary as well as windstorm. And keep the inventory in a safe deposit box away from your home.

How to Avoid Disaster

- If you live in a single-family house in a tornado-prone area and don't have a storm cellar, reinforce some interior room or enclosure as a shelter.
- Have a plan. Be sure everyone in your household knows in advance where to go and what to do in case of a storm warning.

During a tornado:

- Take cover immediately if a warning is broadcast. A tornado warning means a twister has been sighted in or near your community. A tornado's path is unpredictable, so don't take chances that it will miss you.
- Stay calm. Don't attempt to flee in a car or other vehicle. Remember what happened to Dennis and Grace Thorp in Wichita Falls (*see Chapter 1*). A car is no match for the swift, erratic movements of a tornado, which may have a wind speed of more than 200 miles per hour. At least 26 of the 45 persons killed and more than half of those seriously injured in the Wichita Falls tornado in 1979 were attempting to escape the storm in their cars.
- Abandon cars or mobile homes and seek shelter elsewhere. If you're in a car when a tornado approaches, get out and head for the nearest ditch or depression if there is no better shelter available.
- If you're at home or at work, stay inside and away from windows and exterior walls until the storm is over. The safest place in a home during a tornado is in the cellar or basement. If there's no basement, take shelter in a bathroom, a closet or under a heavy piece of furniture (an upside-down sofa would do) on the lowest level.
- If a tornado strikes during school hours, teachers should keep children away from windows and seek shelter in a designated area or in interior hallways. A school building is probably the safest place children could be, so concerned parents should not attempt to go out in the storm to pick up their children at school.

After a tornado:
- Be alert for potential hazards—broken power lines, shattered glass, splintered wood or other sharp protruding objects. Ruptured gas lines can cause explosions from sparks or matches.

How to Handle a Hurricane

Hurricanes needn't concern you unless you live in an Atlantic or Gulf Coast state, or on a Caribbean or Pacific island. But if you do, watch out.

Called the "greatest storms on earth" by the National Weather Service, hurricanes have killed more than 14,000 Americans since 1900 and caused about $8 billion in property damage during the past decade. Although usually limited to the Southern coastal areas of the country, hurricanes have rampaged all the way up to New England and Canada.

While tornadoes pick their way in a narrow if tortuous path, a major hurricane is often 100 miles or more wide and 10 miles high with powerful winds and rain that can sink ships, change coastlines and strip large land areas of buildings, crops, power lines and bridges.

You usually have some notice of a hurricane's approach, through hurricane warning programs sponsored by the insurance industry and others. Newspaper and broadcast advisories usually give area residents information on the storm and on measures to cope with it.

While there is nothing anyone can do to move a house out of the path of a hurricane, there are steps that can be taken to protect life and limb and reduce damage. Here are a few.

In the hours before the storm arrives:

- Leave low-lying areas that may be swept by high tides and waves. If your escape route is over a road likely to be congested or covered by water don't delay evacuation any longer than necessary.
- If you live in a mobile home, check your tie-downs and leave for more substantial shelter. Damage can be kept to a minimum if the mobile home is secured by heavy cable anchored in concrete footing.
- If you own a boat, moor it securely or move it to a designated safe area.
- Board up your windows or protect them with storm shutters. To prevent the occurrence of sudden differentials in air pressure inside and outside the house, some authorities recommend leaving a window slightly open on the side away from the storm.

- Move inside or secure all outdoor objects that might be blown around. This includes garbage cans, garden tools, toys, signs, porch furniture and other seemingly harmless items that can become missiles of destruction in a hurricane. Also make sure that garage doors, awnings and storm shutters are secured. Roof antennas should be removed or lowered.
- Have at least one flashlight in good condition, with extra batteries available.
- If you're in a remote area have a transistor radio handy, and make sure it's working. It may be your only link with the outside world for some time, and your only source of reliable information on the storm.
- Stock a generous supply of drinking water in clean bathtubs, jugs, bottles and cooking utensils in case there is contamination of your water supply. Have extra canned goods and non-perishable foods on hand.
- Make sure your car's gas tank is full, as service stations may shut down. Move your car out of areas that might flood.
- If you depend on any medications, make sure you have an extra supply, as you might not be able to get them during the emergency.
- If you have not been advised to evacuate your area, remain at home—but take no unnecessary risks. Stay away from windows and glass doors where you could be exposed to flying glass. Once the storm has started, stay indoors. Travel is especially dangerous when winds or tides are high.

After the storm:
- Make temporary repairs to prevent further loss. Expenses of such repairs are covered by most insurance policies.
- If you have to go outside, be wary of loose or dangling power lines. Walk or drive carefully. Roads or bridges may be washed out. Debris—even poisonous snakes or insects—may be in your path.
- Check for spoilage of food if power has been out for more than a few hours.
- Make sure tap water is safe before drinking it.

- Cooperate with police and emergency personnel, and with any neighbors who might be in trouble. Report any looters.

Insurance money has helped rebuild whole towns, even cities, after a tornado or hurricane disaster. And insurance companies usually are quick to send in extra personnel to speed the paperwork. So, even though a storm of this kind can be a nerve-shattering experience, it needn't be a financial blow.

How to Keep the Burglar Away

Ancient cities kept flocks of geese at their outer walls to warn of intruders. Smart enemies found ways to circumvent the geese—perhaps by quietly feeding them. And when they did, the geese became part of their own victory feast.

Alarms have come a long way since then. Watch dogs were an improvement, as were mechanical alarms. And the most modern alarm is the sophisticated electronic system. All except the geese are in use today, and you can take your choice of what you want to rely on for burglar protection.

Statistically, you are more likely to have a burglary than a fire or tornado, although your loss probably will be lower unless you keep valuable gems or easily portable art objects in the house. According to the Federal Bureau of Investigation, more than 3 million burglaries were reported in the United States in 1981, and the number keeps rising. Total property loss for the year was in the neighborhood of $2.5 billion, or about $700 per burglary. Depending on where you live, chances are good that your house or apartment will be burglarized at some time. In some high-crime areas, people often report being burglarized repeatedly.

Aside from the property loss, and the sickened feeling most people experience at the forcible invasion of their privacy, burglary also carries with it a danger of violence. Many a householder has lost his life when he either surprised an armed burglar or attempted to do battle with the intruder. Occasionally, newspapers report cases of burglars wiping out several members of a family. That's all the more reason to prevent a would-be burglar from getting inside your house.

There are a lot of things you can do to make your home more secure. First, analyze the problem and size up your particular situation, then take the measures your analysis dictates.

Think about it from the burglar's viewpoint. Knowing about a burglar's three worst enemies—light, time and noise—can help you protect your home from crime. A burglar won't find your home an easy mark if he's forced to work in the light, if he has to take a lot of time breaking in, and if he can't work quietly. Take the time to "case" your house or apartment, just as a burglar would. Ask yourself some questions:

Where is the easiest entry? If trees, shrubs and darkness give a burglar the privacy he would like in order to break in, deprive him of his cover. Trim back trees and shrubs away from doors and windows. Consider installing an outside light that would take away the comfort of darkness.

How can you slow down a burglar? Time, as well as light, is a burglar's enemy. A burglar delayed for four or five minutes is likely to give up in favor of easier pickings. Simple security devices—including such ordinary equipment as nails, screws, padlocks, door and window locks, grates, bars and bolts—can discourage intruders enough to keep them from entering.

What about noise? Noise is that important third enemy of the burglar. So try to make the job of robbing your house a noisy one. With the many types of alarm systems available, deciding just how much home protection you need and can afford is a personal judgment. You can get an elaborate hook-up that will sound inside police or guard headquarters and bring patrol cars or security officers to your house within minutes—perhaps in time to nab thieves red-handed. Sometimes a decal on the front of your house stating that you are so protected will be enough to warn off would-be thieves. Or you can settle for a simple noisemaker alarm that will howl loud enough to wake neighbors (one of whom may call police) if tripped. Or you can purchase something in between.

A large, aggressive dog provides excellent protection—provided he's not so aggressive he'll get you involved in liability suits for biting peaceful visitors or rouse the enmity of your neighbors by constant barking. Even a small, excitable dog which will make noise can be useful. Except for pets so tame they will fawn over anyone, most burglars hate dogs.

Dogs don't always work. In fact, even lions aren't always a deterrent. One man in Texas kept two pet lions and a Doberman pinscher in his house to ward off burglars, but one thief apparently knew how to get along with even the most ferocious animals. He drove a truck up to the house, loaded it up, and took off with practically everything the householder owned—except for the lions and the Doberman. They stayed behind, apparently quite content in the empty house.

One householder's trick was to insert a thin sheet of plywood between the top of his front door and the frame, then pile it high with empty cans and pans that would make an unholy racket if dumped. A corollary would be to place shelves of bric-a-brac beneath windows a burglar might enter, to make a crash when tipped over. Such a noise probably wouldn't summon help, but it could startle a burglar into beating a hasty retreat. At least, the crude alarm would wake the householder himself and give him time to phone police from an upstairs extension.

A man who overdid "security" was an Ohio farmer who became enraged after repeatedly being burglarized by someone who seemed to know when he was going to be away from his house. Swearing to "get" the burglar, he rigged up a shotgun to go off if anyone broke through the door. The plan worked. When the farmer went to town, the burglar entered, tripped the shotgun, and caught the full blast. The burglar turned out to be the farmer's next-door neighbor. Severely injured but not killed outright, the neighbor sued the farmer in civil court, winning a huge judgment that could only be satisfied through seizure of the farmer's house and land. In addition, the farmer faced criminal charges for attempted murder, with the prospect of spending several years in prison.

You might have valuables—a painting, a silver collection, an antique—that are easy to see from outside. If so, rearranging your furnishings might make your home less inviting to criminals.

One good idea is to call your local police department or sheriff's office and ask for someone to make a security survey of your home. Usually this service is provided gladly and free of charge. The surveying officer will point out weaknesses and give you a list of recommendations for making your home more secure.

Here are some specifics to consider:

Doors—Outside doors should be metal or solid hardwood, and at least 1¾ inches thick. Frames should be of equally strong material, and each door should fit its frame snugly. Remember, the best lock in a weak door will not keep out a determined burglar.

Sliding glass doors present a special problem, but there are locks designed for them. A broomstick, or solid bar of wood, in the door channel will keep the door from opening. Of course, the glass itself can be broken or cut. Consider plexiglass for the doors, or an exterior screen with a grate—and a lock.

For identifying visitors, a peephole or wide-angle viewer in the door is safer than a door chain, which an intruder can force once the door is unlatched.

Locks—Deadbolt locks are best. These are locks with a solid bar extending into the frame, rather than a spring latch which can be jimmied open. Some are locked from both sides, others with a key from the outside and with a thumb turn from the inside. The cylinder (where the key is inserted) should be pick-resistant. Ask your hardware dealer for a reputable brand, or buy your locks from a locksmith. For extra nighttime security when you are in the house, you can add a bar on the inside of the door.

Windows—Key locks are available for all types of windows. Double-hung windows can be secured simply by pinning the upper and lower frames together with a nail, which can be removed from the inside. For windows at street level, consider iron gratings. For windows opening onto a fire escape, metal accordion gates can be installed on the inside. But make certain these barriers can be removed easily from the inside in case of fire.

Here are some home security habits you can practice:

- Establish a routine to follow in making certain that doors and windows are locked and alarm systems are turned on.

- Avoid giving information to unidentified telephone callers, or announcing your personal plans in want ads or public notices. Notify police if you see suspicious strangers in your area.
- Handle your keys carefully. Don't carry house keys on a ring bearing your home address, or leave house keys with your car in a commercial parking lot. Have your car keys on a separate ring, or detach the ignition key only. Don't hide your keys in "secret" places outside your home—burglars usually know where to look.

For vacations or weekends away, here are some special tips:

- Leave blinds open in their usual position.
- Have mail and packages picked up by a neighbor, or else held or forwarded by the post office.
- Lower the sound of your telephone bell so its ring can't be heard outside.
- Arrange to have your lawn mowed or your walk shoveled.
- Have newspaper deliveries stopped temporarily.
- Use an automatic timer to turn lights on and off in your living room and bedrooms at appropriate hours.
- Leave a radio playing (not loudly).
- Tell police and dependable neighbors when you will be away, and ask them to keep an eye on your house.

If you should be unlucky enough to confront a burglar in your house:

- Keep calm; don't do or say anything to startle him.
- Get away if you can and call police. A phone extension in your bedroom can come in handy for this.
- Lock yourself in a room if you can't escape from the house.
- If trapped by the burglar, cooperate and do as he tells you. Don't offer resistance or go for a weapon—your life is more valuable than your property. And you might get the worst of it in a shoot-out. Keeping a loaded gun in your house is not advisable. Too often the victim turns out to be a member of your family.

There are ways you can make your property more secure even after you've been burglarized. One of them is through

Operation Identification, a property marking program conducted by many insurance agents and police departments throughout the country. These agencies make available to you an electric stencil or marking stylus, which you use to engrave your Social Security number or other designated identifying number on property items that could be stolen. Later, if you report the theft and the item is recovered, it can be traced to you and returned to your hands. Check with your insurance agent or local police department to see if there is an Operation Identification program where you live. If not, the International Association of Chiefs of Police, in Washington, D.C., operates a program through the mail.

Good security measures for your own home may simply drive the criminal down the street to your neighbor's house, or down the hall to the next apartment. That may trouble you—and it should. Look into some form of cooperative security plan with others on your street or block, or even your neighborhood. Many communities have organized Neighborhood Watch programs in which residents take turns maintaining surveillance of the area.

You can find out about watch programs in your neighborhood by asking your insurance agent or police department. If you want to organize one, they'll be glad to help. Try it. Besides protecting your home, it's also a good way to make friends and get to know your neighbors.

Whatever you do about safety and security, always use your head. Don't get carried away by emotions such as fear or anger. They can backfire on you. And thoughtful planning beforehand is a lot better than hasty improvisation.

A case in point is that of the man who, in fixing his roof, decided, after he had started, to play it "safe" by tying a rope around his waist. Then, throwing the end over the roof peak, he called to his son, and asked him to tie it to something secure. That "something" turned out to be the bumper of the family car, parked in the front driveway. A few minutes later, the man's wife emerged unaware from the house, got into the car, backed it out of the driveway and drove off. She dragged her husband over the roof peak, off the roof to the ground and down the street behind her, until his agonized yells brought her to a stop. Needless to say, he came out of the incident considerably the worse for wear.

CONSUMER TIPS

Prevent Loss

Preventing loss is the best possible insurance and the best possible way of protecting what's yours.

TO PREVENT **FIRE-RELATED LOSSES**:
- ☐ Eliminate potential fire hazards in and around your home.
- ☐ Install smoke detectors.
- ☐ Develop and practice a family escape plan.
- ☐ Use woodstoves, heaters and fireplaces carefully.

TO MIMIMIZE OR PREVENT **WINDSTORM-RELATED LOSSES**:
- ☐ Make preparations for tornadoes or hurricanes **before** they arrive.
- ☐ Heed evacuation warnings and other safety directions **during** and **following** a major windstorm.
- ☐ Make needed temporary repairs and observe special safety precautions **after** the storm subsides.
- ☐ Contact your insurance representative as soon as possible.

TO PREVENT **BURGLARY-RELATED LOSSES**:
- ☐ Slow burglars down with locks, bars, grates and other simple devices.
- ☐ Deprive burglars of privacy by trimming trees and shrubs and lighting entryways.
- ☐ Make robbing your home a noisy job using alarms or a watchdog.
- ☐ Remove valuables from view by outsiders.
- ☐ Check with your insurance agent or local police department to see if there is an Operation Identification program where you live, or secure the necessary equipment for marking your valuables through the International Association of Chiefs of Police.
- ☐ Take an active part in preventing crime in your community. Help organize a block association or Neighborhood Watch program. Ask your insurance agent or local police for help.
- ☐ **Do** use your head to organize and plan for safety and security.
- ☐ **Don't** let your emotions get involved.

PART TWO

Protecting What's Yours On the Road

CHAPTER 10

YOUR AUTO INSURANCE

Of all the hazards and perils of life that threaten your fortune and possessions, probably none is more constant nor greater than the chance of having an accident with your car. Motor vehicles maim, they kill, they destroy houses, wipe out families, and even disrupt businesses. Even the most innocent of trips can end in an accident that triggers hundreds of thousands of dollars in claims against you or your estate. Sometimes the results can be quite surprising—and staggering.

For instance, one motorist accidentally crowded a flat-bed truck off a highway. Although the truck wasn't damaged, it was carrying an intricate milling machine, and the machine suffered several thousands of dollars worth of damage. But that was just the start. The company that owned the machine filed a claim against the motorist, alleging that the time the machine was out of service for repairs would cost it in lost business, and the company demanded $100,000.

Fortunately automobile accidents should not wipe us out, because we have automobile insurance. Unfortunately some of us don't have as much as we should. For instance, that motorist who damaged the machine had only the required minimum coverage for property damage liability—$5,000.

Naturally auto insurance didn't exist when the first cars were being built in the United States. Nevertheless driving presented substantial risks, especially if a car startled a horse and caused a runaway. The first automobile policy was issued to Gilbert J. Loomis in Westfield, Mass., in 1897. Loomis, a talented mechanic who later founded the Speedwell Motor Car Co. in Dayton, had built his own one-cylinder car. He bought a $1,000 policy for $7.50, which, when compared with today's premium for $1 million of auto liability and "umbrella" protection, was astronomically expensive.

Soon after he had built his car, Loomis received a registered letter from the town's three selectmen forbidding him ever to operate "that thing" on the town's streets. There is no record of whether Loomis wheedled the selectmen into letting him drive so long as he had liability insurance, but if that was the case—and it likely was—it marked the first time a government ever set up a requirement that a driver be financially responsible for any liability he might incur, a proviso that many states now require. That means if you want to drive, you must be able to pay for any damages or injuries you cause out of your own pocket or with insurance. Obviously one big accident could clean out your pocket and take your shirt as well. That's why automobile insurance is so crucial.

Just suppose you're driving late one night, your eyes get heavy, you nod, and in that moment the car veers across the center line and hits an oncoming car head on. Both cars are totaled, and the other driver has hit his head on the windshield, breaking his neck and paralyzing him from the neck down. No matter how careful a driver you might be, the accident was your fault, and you stand liable to pay the other man's bills, plus nursing care, and his living costs for as long as these expenses continue—even a lifetime. A Florida jury awarded a 21-year-old college student more than $2 million for similar injuries. If you think that's steep, actuarial tables show that the cost of caring for that man over the rest of his life will probaby total $3.4 million.

Obviously that kind of liability can bankrupt you unless you have insurance with extra high limits. And your need for stepped-up protection is growing, because juries are awarding increasingly larger judgments. Moreover, people are more inclined to sue these days. A Michigan man who drove down the wrong side of the road and killed a couple when he crashed into their car head on, filed for damages from the couple's estate on the grounds that the other driver had had time enough to get out of the way.

If it happens that the other driver was at fault or there's some question about who's responsible, you will need expert lawyers and you may face a lengthy trial in order to protect yourself. But that again is where your liability insurance can save you. The insurance company will pay for your defense, even providing the lawyers. If they are not successful in defending you, the insurance company will pay the judgment against you— up to the limits of your policy.

The big question everyone asks these days is how much coverage a driver should have. Obviously you should get more liability coverage than the minimum required by most states, which runs around $25,000 to $50,000.

Fortunately extra liability protection is relatively cheap. For only a few extra dollars you can raise your coverage limits to $300,000 or $500,000, or even more. And you can buy separate umbrella liability coverage for almost unlimited amounts. The reason you get so much protection for so little extra is that these astronomical awards are relatively rare. Nevertheless, you could be one of these rare cases.

You probably can offset some, or even all, of the additional cost by tailoring your other coverages to fit your precise needs. While liability protection is the heart of every auto insurance policy, a number of other coverages are available as well. The trick is to add only those coverages you really need and avoid adding extras that duplicate your other insurance or which you run little risk of ever needing.

The Auto Insurance Package

Auto insurance policies have become pretty well standardized over the years since Loomis took out liability coverage on his one-cylinder flivver. Many insurance companies apply their own distinctive names and special features to their policies. And, despite the fact some of the more cumbersome, traditionally structured policies are still being offered in some states today, the simplified, more standardized insurance packages are now used over most of the country. Most motorists buy this standardized package. These packages offer the same six basic types of coverage: bodily injury liability, property damage liability, medical payments coverage, uninsured and underinsured motorists coverage, collision insurance, and comprehensive coverage. And some states have laid on top of all this what is known as "no-fault" coverage, which is explained in the next chapter. If you are to spend your premium dollars more wisely and not be bludgeoned by a lawsuit some day, you need to understand the basics of what these coverages will do for you.

Bodily Injury Liability

If you should accidentally kill or injure someone, you may be held liable for his or her hospital and medical bills and such other costs as funeral expenses, the costs of rehabilitation and long-term nursing care, and, in some cases, the victim's lost earnings. The court might award the other person an additional sum for mental anguish, or pain and suffering. Your bodily injury liability insurance is designed to pay for all that, assuming you're covered. In some states, the court could award the other party punitive damages to punish you for your negligence.

Property Damage Liability

This covers your liability if you damage someone else's property. Normally it involves a car, but sometimes you can damage other types of property, such as someone's personal belongings, or a house or factory, or trees and fences. If you run into a highway sign, the state will make you pay for it, but your property damage liability insurance will cover the cost. Property liability can even cover some of the other person's expenses from the accident, such as if he or she must rent a substitute car.

Liability coverage is extremely important, and drivers often fail to cover themselves sufficiently. One driver sideswiped a dairy tank truck, causing it to turn over and spill its entire cargo of fresh milk. Damage to the truck and the loss of the milk totaled $12,300, but the driver was covered for only the minimum requirement in that state—$5,000—and he had to make up the $7,300 difference out of his own pocket. In California a car hit a fire hydrant, and water shot up onto the flat roof of a nearby factory at a rate of four tons every minute. It wasn't long before the factory roof collapsed, causing about $15,000 in damage. Again the car's owner was insured for only $5,000.

Yet extra liability coverage for $100,000 or $300,000 adds only a fraction to the total cost of your auto insurance. So it really is a good idea to insure for extra coverage.

Medical Payments Coverage

To cover the doctor and hospital bills that you and your passengers incur, there is medical payments coverage. Your insurance company protects you regardless of who caused the accident. That means you can have your bills taken care of without having to go to court or file a claim against the other driver. If the other driver was at fault, your insurance company has the right to collect its costs back from that person or the other insurance company.

Just how much your insurance company will pay depends on how much coverage you decide to take out when you buy the insurance. It can range from $500 up for each person in your car, and for each accident you have. But $500 is dangerously low considering today's high medical costs. If you don't have an extremely good health insurance policy, you'd better get considerably more than that.

You don't have to have what you normally think of as a traffic accident to be covered by medical payments insurance. You can be covered if you accidently slam a car door on your finger. And medical payments covers you and your family if you're pedestrians and are hit by a car or even a moped.

One thing you should remember. Generally, medical payments coverage applies only to non-paying passengers. If you are carrying anyone who is paying you for the ride, you'd better check with your insurance agent and see whether you need extra coverage. Otherwise, it might cost you if there's an accident.

There are a few situations where the medical payments coverage doesn't protect you. If you're riding a motorcycle or bicycle or any other two- or three-wheel vehicle you'll need other insurance. And if you're driving your own car on business and are entitled to workers' compensation you won't be covered.

Should you be injured while riding in someone else's vehicle, their insurance will pay your medical bills up to the limits of their coverage. If your bills total more than that, your own policy will take over when the other coverage is used up and pay up to its limits.

The cost of medical payments coverage is quite low and accounts for a surprisingly small portion of your auto insurance premium. So it is something worth having. For a few added dollars, some policies offer death and disability benefits similar to those provided in life and health insurance policies except they cover only victims of highway accidents. If you have life and disability coverage in other policies, you probably don't need this with your auto insurance.

Uninsured Motorists Coverage

If you're hit by some irresponsible driver who doesn't have liability insurance, or any assets worth mentioning to collect from, there's a provision in your auto insurance that will cover you. Although it is a minor part of your insurance package, this feature could be of utmost importance to you some day.

If you have uninsured motorists coverage, your own insurance company will protect you from loss—which could be considerable—if:

1. You are the victim of an accident caused by a motorist without insurance.
2. Or you are a hit-and-run victim.

And, under some policies, you are covered if you're the victim of a motorist who has coverage with an insurance company that has gone broke.

This coverage applies to you, to members of your family, and to any non-paying passengers in your car. Usually it covers only claims involving bodily injuries, but in some states you can get additional coverage for damage to your car. Uninsured motorists coverage can be a life-saver in a major accident, but one word of caution. Just because the other driver wasn't insured does not mean you can collect automatically from your own insurance company. You have to prove that the accident was the other driver's fault. It's easy to blame a crushed fender on a hit-and-run driver who's not around to defend himself. But it's not so easy to prove that's what happened, especially since some motorists try to use this ploy to get dents fixed that they themselves were responsible for.

As a result you could wind up in a dispute with your insurance company. In case that does happen, the policy sets up an arbitration procedure you would follow should you and the company fail to agree either on the amount of the damages or whether you're entitled to them. Arbitration works the same here as with homeowners insurance. You and the company each choose an arbitrator and they choose a third. The decision of two of the three arbitrators is binding.

There are a couple of other exceptions you should know about. Quite naturally uninsured motorists coverage does not cover you if the other car in the accident is owned by you or another member of your family but not covered under your policy. And it doesn't cover you if a government-owned vehicle or an off-road vehicle, such as a tractor, causes the accident.

Underinsured Motorists Coverage

Usually this is an optional coverage—and it's available only in some states—but it's one you ought to consider seriously. Even if the other driver is insured he might not have more than the minimum coverage, and you still could be buffeted by bills after his insurance company has paid off. In a situation like that, your underinsured motorists coverage will make up the difference, up to the limits of your policy.

In certain cases this could be quite valuable coverage. But you don't want to overinsure yourself if you are already covered for this sort of problem in your non-auto policies. Get your agent's advice, but before you even see him check and see how much medical and disability coverage you already have under your other policies. If you are not protected by other insurance, then you should consider this as part of your auto coverage.

Collision Insurance

One of the cornerstones of almost every auto policy, collision insurance pays for damage to your own car. Your collision coverage will protect you whether your car is damaged hitting another vehicle, or some stationary object.

Collision coverage can account for a large part of your auto insurance premium. In fact, if your car is more than five years old perhaps you should consider dropping your collision coverage, because the cost might not be worth what you'll get back.

A little arithmetic will tell you why. An older car costs just about as much to repair as a new one. Yet your premium is based on what your car cost when it was new, and if it's more than five years old you can bet you didn't pay anywhere near what a new one costs today. So your insurance company isn't going to pay you more for collision damages than the "book," or current market, value of your car. If damages total more than that, the company declares the car a total loss and pays you the book value, rather than spend more than the car is worth to fix it. Since a car depreciates in five years to about a third of its original price, your payoff on a car that's at least five years old could be quite skimpy, considering the premiums you'll be paying year after year.

If you have a new car or an expensive car whose value is still relatively high, you ought to have collision insurance. The premium is minuscule compared with the money you can lose if you're in an accident, especially at today's repair prices. (It costs more than three times the price of a new car to replace a car part by part.) Of course if your car is an antique you should have special collision coverage that your agent easily can get for you.

One way you can hold down the cost of your collision coverage is to take a higher deductible, which is the amount of each loss that you will pay before the insurance comes to the rescue. Usually these deductibles run from $50 to $500. The higher your deductible, the lower your premium. Although prices vary, you can expect to save about 35 percent on your collision coverage if you raise your deductible from $200 to $500. Don't forget, if you increase your deductible you'll have to pay a higher portion of the cost to repair your car if there is an accident. But so long as you have the money to cover that extra cost you'll be way ahead with the higher deductible.

Besides the obvious fact that it protects you from repair costs that could run into the thousands, an important advantage of collision coverage is the way it assures you of getting the money right away to repair your car. For instance, if your car is damaged in an accident that is the fault of another driver, you won't necessarily have to wait to collect from the other driver's insurer. With collision, your own company will pay for the damages, less the amount of the deductible of course. Your own company will then take the necessary steps to collect from the other driver or his insurer, and if it is successful you may even get back the amount of your deductible.

Comprehensive Coverage

If something should happen to your car other than an accident, your comprehensive, or "other than collision," coverage will protect you. For instance, if the car is stolen, your insurance company will reimburse you for the value of the car and will even pay a reasonable amount for substitute transportation, such as a rental car, perhaps for as long as 30 days.

As its name implies, comprehensive can cover a wide range of circumstances. For example, one Eastern woman lent her car to her boyfriend with the understanding he would drive it across the country and meet her on the West Coast to be married. But once the boyfriend set off, he disappeared. After several months and no word from him, the woman reluctantly concluded she had been jilted—and her car stolen. Her comprehensive insurance covered the car, but not her broken heart.

Comprehensive also pays for damage to your car from virtually any cause other than collision. For instance:

1. Fire.
2. Flood.
3. Falling objects.
4. Encounters with animals (such as the bird that crashes into your windshield or the deer that runs into your path).
5. Vandalism or malicious mischief.

Insurance companies usually interpret comprehensive coverage quite liberally. One woman collected after her car was scratched by a pack of dogs. A man locked his dog in the family car while he went off for a few minutes. The dog expressed his objection to being left behind by completely tearing up the car's upholstery. But the man's comprehensive took care of the loss.

Some policies also extend comprehensive coverage to include rented or borrowed cars. And, although comprehensive is designed to pay for damages to your car rather than other property you might be carrying with you, some policies provide limited coverage for your clothes and other personal effects if they are in the car and the loss is caused by fire or lightning. But such personal possessions as CB radios, sound recording equipment, telephones and scanners must be installed permanently in the car if they are to be covered.

Traditionally, broken glass has been covered by comprehensive rather than collision insurance. Nevertheless, insurers usually have been willing to pay for broken glass under the collision coverage if the policyholder did not have comprehensive.

One hitch encountered by some drivers who carry both collision and comprehensive is double deductibles. If glass is broken and the car's body damaged in the same accident, the driver has been stuck with having to bear the cost of two deductibles. To accommodate this problem, many insurance companies are now giving the car owner the option to collect for all damage from an accident—including broken glass—under the collision coverage. So, shop around before you buy and see what's offered.

You also should examine what your own exposures to loss would be if you didn't have comprehensive coverage. For example, if you drive in wooded country you probably need to be covered against collisions with deer or other wild animals. If you face that kind of risk, comprehensive coverage would be well worth the cost.

Miscellaneous Coverages

While such coverages as medical payments and comprehensive are basic to your auto policy, there are other coverages as well that are optional. They are there so that you can tailor your policy to your needs. Don't load all these extras on just to have them. Watch your premium dollars and think carefully about which coverages you'll really need and whether they will duplicate other insurance you already have.

Towing and Labor Costs—This coverage is usually available as an option. It pays up to a fixed amount, typically about $25, for any towing and labor charges, whether your car is disabled in an accident or just breaks down—even if it's in your own driveway. Labor performed in a garage isn't covered, only work done at the scene of the breakdown. If you are a member of an auto club, you already have this protection. Or, you may feel it is a cost you could absorb out-of-pocket. If so, you could save by skipping this option.

Towing coverage also wouldn't help you very much in a really difficult situation. For example, an Ohio man ran up a $900 towing bill when he ran off the road, hit some trees and landed in a Mississippi swamp. Fortunately, damage to his vehicle was fully covered by his collision insurance.

Rental Reimbursement—Except when your car is stolen and you have comprehensive coverage your auto insurance won't pay for a rental car—which you may need while your own car is in the shop after an accident. But many companies offer an optional extra coverage called rental reimbursement. If your car is laid up for more than 24 hours, your insurer will pay up to $15 a day toward the cost of a rental car. Remember, you'll have to keep your receipts. The coverage isn't intended to pay for transportation such as buses, trains and the like. This policy endorsement also is available if your car is stolen, and may be used to increase the amount provided for car rentals under your comprehensive coverage.

What If It's a Whatchamacallit?

Besides a car, do you drive a motorcycle? A motor scooter or a moped? A motor home? A dune buggy? A three-wheeled dingus with a plastic rain shield, that you use only to get you to the station in the morning? A contraption that your teenage son built out of bicycle parts, lumber scraps and a one-cylinder engine?

When you're out in traffic in any other motorized vehicle, you have the same exposure to a crack-up as you'd have with a car. But insurance companies don't consider any of those other machines to be automobiles, so your regular auto policy won't protect you while you're using one.

No real problem. You can get essentially the same coverages for your motor-bike or your whatsit through an endorsement (addition) to your auto policy. If you never have anybody with you except members of your family, you probably won't need coverage for injuries to other passengers. Some states permit you to exclude that.

Some companies offer six-month or nine-month policies to people who live in colder climates and retire their motorcycles during the winter months.

Bear in mind that if you have a motorcycle or a moped, certain states require that you have liability insurance. And it's a good idea, whether required or not. Bicycles are covered by the liability coverage in your homeowners policy.

You can get special insurance coverage for your golf cart if you have one, but if you have homeowners insurance, you probably won't need it unless you regularly drive the cart to and from the golf course on a public road. When you use the cart on the golf course, you're protected by your homeowners policy. That protection includes your liability to others in case you run into something or someone, or overturn and injure a passenger.

Insurers make a special case for snowmobiles. Most companies will insure them for use off your property through an endorsement to either a homeowners or an auto policy. The homeowners endorsement is more restrictive, in that it includes only liability and limited medical payments coverages and can't be used to insure snowmobiles that are subject to motor vehicle registration, as required in several states.

The endorsement to the auto policy provides a full range of coverages, including uninsured motorists and physical damage insurance, and may be used in states which require snowmobiles to be registered. It would depend pretty much on your use of the snowmobile and a comparison of the added premiums, which type of endorsement would be the better buy for you.

If you've got some other kind of motorized vehicle parked in your garage that doesn't seem to fit any of these categories, talk to your agent or company. They'll probably be able to figure out a way to cover it.

Not to cover it, or just to assume you're covered under your standard policies, could be expensive. You could find you own a legal boobytrap.

What the Law Requires

You can find both your property and your freedom in jeopardy if you fail to obey the legal requirements that the various states, Canada and Mexico impose on drivers. More than half the states have laws requiring their registered car owners to carry insurance to pay for potential injuries to others or damage to others' property.* (As an alternative, some states require the car owner to post security.) Penalties for failure to comply include fines, loss of registration or driver's license, and even jail sentences.

Even residents of states that don't require liability insurance need it just to drive in most other states.

Most states also have reciprocal arrangements, meaning if your license is picked up elsewhere you may stand the chance of losing both your license and your registration in your home state.

---

*As of January 1983, you're required to have auto liability insurance if you own a car and register it in California, Colorado, Connecticut, Delaware, Georgia, Hawaii, Idaho, Kansas, Kentucky, Louisiana, Maryland, Massachusetts, Michigan, Minnesota, Montana, Nevada, New Jersey, New York, North Carolina, North Dakota, Oklahoma, Oregon, Pennsylvania, South Carolina, Texas, Utah, West Virginia or Wyoming.

The laws apply to motorcycles as well as private passenger cars in all of those states except Colorado, Connecticut, Georgia, Montana, New Jersey, North Dakota and Utah. Mopeds are exempt in Connecticut, Michigan, New Jersey and North Dakota; farm vehicles in Connecticut, Kansas, Montana and New Jersey.

Generally, the laws of most states and the provinces of Canada say you may be required to produce evidence of financial responsibility up to specified amounts if:

- You're involved in an accident resulting in an injury or property damage in excess of a stated amount (ranging from $50 to $500);
- You're convicted of a serious driving offense—such as drunk driving, reckless driving or hit-and-run driving—or a series of offenses;
- You fail to pay a judgment growing out of an automobile accident.

If you have auto liability insurance, it will serve to meet the requirements of the financial responsibility laws of the various states and provinces, even though the minimum limits elsewhere exceed the amount of your coverage.

Most policies provide that if you have an accident in another state or in Canada and you don't have the amount of coverage that's required there, your insurer will meet any liability claim up to the minimum financial responsibility limits of the other jurisdiction. Many companies provide you with that same added protection for an accident in a state that requires out-of-state drivers to have liability or no-fault auto insurance.

If you have no insurance and you run afoul of a financial responsibility law, you may have to post some other form of security such as a bond or a cash deposit. You also may have to show proof of your financial responsibility for a future period which might be as long as three years.

In view of this mix of insurance requirements from place to place, it's a good idea to carry with you wherever you drive evidence that you have auto liability coverage. It may save you some embarrassment—or worse—if you have an accident or are stopped by the police for any reason.

Some states say you must have proof of insurance with you. Give your insurance representative a call if you have any question as to what kind of proof you need to keep in your car.

Visiting Neighboring Countries

Virtually all auto policies issued in the United States cover accidents anywhere in this country, U.S. territories and possessions, and Canada. But if you're heading for Canada, especially if you'll be traveling extensively there, you can guard against trouble by (1) having insurance and (2) obtaining from your insurer an insurance identification card that will serve as evidence you're covered.

If you're planning to drive into Canada or Mexico, talk with your insurance representative before you start. Here's why:

In most of the Canadian provinces, you may be required to provide evidence of insurance if you have an accident. If you can't, you'll face the prospect of losing your driving privileges and having your car impounded.

Don't take the wrong car. The "Canada Non-Resident Inter-Province Motor Vehicle Liability Insurance" card describes the car you'll be driving and gives details of your coverage. It also warns you that if you don't have the insurance the card says you have, you're liable to "a heavy fine and/or imprisonment" and that your license may be suspended.

Mexico can be downright ruthless. You can go to jail if you're caught without adequate and acceptable insurance after an accident. In addition, your car may be impounded and your license may be picked up until you satisfy the authorities of your ability to pay the damages.

One important fact you should remember: Your American auto insurance is no good in Mexico. Under Mexican law you have to buy insurance from an authorized Mexican insurer. It's generally available at border crossing points.

Because of the stringent penalties, driving in Mexico without insurance is risky. If you're stopped for any reason, you may also be required to show a valid driver's license, proof of ownership of the car and a tourist card. You can obtain a tourist card at the border, from the Mexican consul or from an office of the Mexican Bureau of Tourism.

Questions That Arise

What's a collision? This question has been battled out in many a court. In one case, for example, fire under the dashboard caused a driver to lose control and run into another car. Although there was a collision, a California court decided the fire was the cause of the accident, and that the cost of repairing that driver's car should be paid under the comprehensive part of the policy. Such a finding could make a difference if the car owner happened to be carrying collision but not comprehensive coverage, or if the amounts of the deductibles were different under the two coverages.

Or, suppose you lose control of your car, and it plunges over an embankment into a lake. Although most of the damage may be caused by the soaking, a number of courts have ruled such an accident qualifies as a collision—in this case, with the lake.

Such decisions are not necessarily binding on other courts in other cases. Every claim is considered on its own merits, depending on exactly what happened—and why. It's just about impossible to nail down in advance every possible complication. Someone—either an insurance claims person or a court—may have to make a decision.

Will you be able to replace your car? If your car is a total loss, and your insurer pays you its "book value," you may find it difficult to find another car just like it on a used-car lot at that price. So you have a couple of choices:

1. You can buy the best car you can find for the money—and since you may be under time pressure because of your immediate need, you may end up feeling you didn't get a very good exchange.
2. You can add some money of your own to the insurance settlement and buy a better car—or use the money as down payment on a new car.

Any way you work it, you're likely to feel you've been put in a bind, perhaps through no fault of your own. This has been recognized as a problem, and certain companies have recently come up with a "replacement" policy which for an extra premium guarantees to buy you a new car, or a newer one, in case yours is wrecked beyond reasonable cost of repair. The policies may not be available everywhere, so you should ask an agent or company representative about this coverage if you are interested.

What isn't covered by your policy? You should be aware there are certain circumstances which your auto insurance doesn't cover. Most of these conditions are spelled out in your policy as "exclusions."

Under your liability insurance, you may find you are on your own, for example, if you or anyone in your family:

- Cause an accident intentionally;
- Use someone else's car without permission, or without at least a reasonable basis for believing you're entitled to use it;
- Have an accident with a vehicle not belonging to you, that you are using while engaged in the business of selling, servicing, repairing, storing or parking cars. In such a case, it's best to make sure you are covered by the insurance carried by the company you're working for.

Your auto policy doesn't cover motorcycles, or any other vehicle with fewer than four wheels—bicycles, for example. Should you be hit by another vehicle while bicycling, you would be covered under the medical payments portion of your own auto policy. Should you run down somebody with your bike, you wouldn't be covered by your auto liability insurance, but your homeowners liability would cover.

Your comprehensive coverage usually will exclude damage that is part of widespread and unavoidable damage resulting from violent man-made events—such as radioactive contamination, blast from nuclear weapons, war (civil or otherwise), insurrection, rebellion or revolution. You are covered against riot and civil commotion, along with explosions and violent acts of nature: earthquake, windstorm, hail, water or flood.

Who's covered and in what car? All members of your family who live with you generally are covered under your auto liability insurance. That includes your spouse and children, blood relatives, adopted children, in-laws, wards and foster children, whether they're driving your car or someone else's. They have the same protection as you, and they're subject to the same exclusions. But your insurance company will want to know which members of your household will be driving your cars.

You can cover more than one car at a time. But you'll have to give your insurance company the information for listing each car, trailer or other vehicle on your policy. Other vehicles on your policy may include panel trucks, pickup trucks or vans not used in a business. Generally, you'll be charged a lower premium for each additional vehicle than for your first car, if you insure them with the same company. If you trade cars, your insurance will be continued, but you should let your insurer know about the change of vehicles. You'll generally have a 30-day grace period in which to notify your insurance company about new or additional vehicles.

This provision saved one car owner a few dollars. A Nebraska man, who had full coverage on his new Chevy Citation, including collision and comprehensive, bought a second car, an old Ford. Before he had a chance to inform his agent about the additional car, he had an accident and totaled the Ford. When he called in the claim, the agent told him he was lucky, as ordinarily he would not have recommended collision coverage on a $300, 10-year-old car. But because the Chevy policy automatically gave him full coverage for 30 days, he collected $200 ($300 less his $100 deductible).

Other cars in your family are another story. Except during the grace period, you definitely will not be covered if you're driving a car of your own that's not listed on your policy. If a live-in relative has a separate car, things get a little complicated. If you or your spouse drive it, you're covered under your policy. But your policy doesn't protect other family members while they're driving the relative's car.

Usually, you're not covered if you're driving a car that's not yours, but is "furnished and available" to you on a regular basis. For example, if you drive a company car regularly, your own liability insurance probably won't protect you while you're using it. Ask your employer if the company insurance covers you, or see your own insurance agent or company about special coverage.

What if you lend your car to a friend or someone you work with? Under most circumstances, other people using your car with your permission are covered by your liability insurance. Yes, the borrower normally would be covered by his own policy. But many states have laws which say the owner of the vehicle may also be held liable if a borrower's negligence causes an accident.

Laws or no laws, chances are that a badly injured victim will file a claim against you as the owner as well as against the friend who was driving your car. And there's always the chance your friend or associate will have no insurance, or not enough insurance or assets to satisfy the claim. Always be careful about lending your car. You could leave yourself wide open to a claim if you lend to a minor or someone you know to be a poor risk, such as a person with a record of drunk driving convictions.

In some cases, your auto liability insurance would even protect your employer or an organization you're associated with. Most policies would protect your employer, for example, if you or a fellow worker were to have an accident while using your car on company business, or while driving a borrowed or rented car on your firm's behalf.

Suppose, instead of owning your car, you rent or lease it? Some policies say in so many words that if you lease your car under a written agreement for a period of six months or more, the car is considered to be owned by you for insurance purposes. If you do plan to lease, look at your particular policy to be sure. In any event be sure you're covered. If you plan to rent or lease a car for more than a month, talk with your agent. If necessary you can add an "endorsement" overriding the "regular use" exclusion.

What about rented cars driven for a brief period? No problem. You're protected under your own liability coverage. Collision insurance is included in the rental fee, with a fairly high deductible. You may be asked to purchase extra insurance to cover the deductible amount. But if you already have collision coverage on your regular policy, that shouldn't be necessary as most companies will cover you for the difference in deductibles. To be sure, pick up the phone and check with your own insurer.

What if passengers in your car are injured? Let's assume that you and several friends are headed for the beach in your car, you have an accident, and you're all hurt. Now what?

Normally, your liability insurance will cover any injuries to your passengers, providing the accident was your fault. If it wasn't, your medical payments coverage will pay, except in states that have "no-fault" insurance, where the injuries would be handled under that form of coverage. But some states have laws called "guest statutes" specifically barring legal liability to non-paying guests unless the accident was caused by your gross negligence.

Gross negligence involves such carelessness as driving while under the influence of alcohol or drugs, or showing off how fast your car can go. You'd probably be held grossly negligent if you drove knowing that your car had bad brakes or some other serious mechanical defect.

Guest statutes, as of late 1982, were in effect in Alabama, Arkansas, Colorado, Delaware, Illinois, Indiana, Iowa, Massachusetts, Nebraska, Oregon, South Carolina, Texas, Utah and Virginia.

CONSUMER TIPS

In Search of the Best

Get the best value for your auto insurance dollars. It's basic to protecting what's yours. Be sure you select the coverages best suited to your needs—it's a big part of maximizing the value of your insurance.

Here are some tips on the basic types of auto insurance coverages and what they can mean to you and your pocketbook.

LIABILITY INSURANCE

Liability insurance is the most important type of auto coverage. There are two kinds of auto liability insurance:

- ☐ Bodily injury liability provides money to pay claims against you and the cost of your legal defense if your car injures or kills someone.
- ☐ Property damage liability provides money to pay claims and defense costs if your car damages the property of others.

You should buy more liability coverage than the minimum required by your state's financial responsibility law or compulsory insurance law. The extra protection you get for a few extra dollars would prove to be a real bargain should you be involved in an accident.

"Umbrella" liability coverage also is worth considering if your assets are substantial.

MEDICAL PAYMENTS

Medical payments insurance pays medical expenses resulting from accidental injuries and covers you, your family, and other passengers in your car.

- ☐ If you don't have a very good health insurance policy, you'd better buy more medical payments coverage than the average policy provides—$500 per person per accident.
- ☐ Passengers in your car who pay you for the ride won't be covered under medical payments insurance. If you transport paying passengers, talk with your agent about extra coverage.
- ☐ The cost of medical payments coverage accounts for a very small portion of your auto insurance premium, so it's well worth having.
- ☐ Death and disability benefits offered by your auto insurance company may compare closely with coverages in your life and health insurance policies. Avoid duplicating coverages.

Your Auto Insurance

UNINSURED AND UNDERINSURED MOTORISTS COVERAGE

Uninsured motorists protection pays for injuries caused by an uninsured or hit-and-run driver.

Underinsured motorists protection is an optional coverage which makes up the difference between the other driver's coverage and the amount of loss he causes you.

- ☐ Check to see how much medical and disability coverage you already have under your health insurance policy.
- ☐ Seek your agent's advice.
- ☐ Make an informed decision about uninsured and underinsured motorists protection.

COLLISION INSURANCE

Collision insurance pays for damage to your car resulting from a collision or from overturning.

- ☐ This coverage pays for damage to your own car.
- ☐ Collision coverage can account for a large part of your insurance premium.
- ☐ If your car is more than five years old, you should consider dropping collision coverage.
- ☐ If your car is new, you should have collision coverage.

COMPREHENSIVE COVERAGE

Comprehensive physical damage insurance pays for damages when your car is stolen or damaged by fire, flooding, hail or other perils, but not when it is damaged in a collision or when overturned.

- ☐ You should base your decision about comprehensive coverage on the age and value of your car and on your own exposure to these types of losses.

Use This Checklist To:

- ☐ note which types of auto insurance coverages you have an in what amounts, and
- ☐ review your understanding of each type.

TYPE OF COVERAGE	AMOUNT OF COVERAGE
☐ Bodily Injury Liabiity	_____
☐ Property Damage Liability	_____
☐ Medical Payments	_____
☐ Uninsured Motorists Insurance	_____
☐ Underinsured Motorists Insurance	_____
☐ Collision Coverage	_____
☐ Comprehensive Coverage	_____

There are two optional coverages to consider:

- ☐ Towing and labor costs

    Decide whether you need this type of coverage by asking yourself two questions.

    - ☐ Does my Auto Club membership provide money for towing and labor?
    - ☐ Can I afford to absorb these costs out-of-pocket? Remember that neither duplicating coverages nor insuring for small losses is ever a good idea—not a smart way to protect what's yours.

- ☐ Rental reimbursement. If you depend on your car for getting to and from work, this one may be worthwhile. If public transportation is available, forget about this endorsement.

Remember your regular auto policy applies **only** to your auto—not to your motorcycle, moped, snowmobile or whatever. If you ride or drive:

A motorcycle or moped:
- ☐ Ask your agent whether liability insurance is required in your state.
- ☐ Buy some liability coverage even if it isn't required.
- ☐ Ask about six- or nine-month policies if you use your vehicle only in warmer months.

A bicycle:
- ☐ Liability coverage is provided by your homeowners policy.

A golf cart:
- ☐ On the road, you'll need a special liability endorsement.
- ☐ Only on the golf course, your liability protection is included in your homeowners policy.

A snowmobile:
- ☐ On your property, you're covered in your homeowners policy.
- ☐ Off your property, get an endorsement either to your homeowners or auto policy.

What endorsements might you need to consider?

☐ Motorbike?	☐ Motor home?
☐ Golf cart?	☐ Dune buggy?
☐ Snowmobile?	☐ Other?

Keep in your car an insurance card or some proof you have auto liability insurance.

When you plan to drive in Canada:
- ☐ Have a talk with your insurance representative.
- ☐ Secure a "Canada Non-Resident Inter-Province Motor Vehicle Liability Insurance Card"

When you plan to drive in Mexico:
- ☐ Purchase a special policy at the border.
- ☐ Secure a tourist card.
- ☐ Take something with you to prove ownership of the car.

Use these strategies to save $ on auto insurance:
- ☐ See to it that your coverage is tailored to fit your personal needs.
- ☐ Add to your liabilty coverage only the additional types of insurance you require for protecting what's yours.
- ☐ Avoid duplicating coverages provided in your other insurance policies.
- ☐ Reject coverages you run little risk of needing.
- ☐ Take the highest deductibles you can afford on collision and comprehensive insurance.
- ☐ Understand your coverages and what they do—and do not—provide.
- ☐ Phone your agent to ask questions about anything you don't understand.

CHAPTER 11

ALL ABOUT NO-FAULT

Although auto insurance originated as a protection for the driver who causes an accident rather than a compensation system for victims, liability insurance does take care of the victim, too, particularly if the driver's fault is beyond question and there is no debate over the amount of damages. Usually the driver's insurance company accepts responsibility, settles the case out of court, and pays the bills.

But what if there's a question or argument over who was at fault? Suppose in addition to direct medical bills, you, the victim, claim large sums for disability or for "pain and suffering." Such claims may be open to challenge.

If the claims are substantial and fault is not clear, the case probably will go to court, and that means inevitable delays of months or years before it comes to trial. The outcome can depend on the reactions of a jury, the quality of the evidence and the talents of the opposing attorneys—with the burden on the claimant to prove his case. And the verdict, if it comes, can range from niggardly to over-generous, depending on the jury's emotions. Moreover, you'll never see it all. About one-third of that money will go to your lawyer.

For decades critics have called for an end to such procedures in auto accidents, urging their replacement with a system that compensates accident victims directly—regardless of who was at fault. Dissatisfaction was voiced publicly, not only by purchasers of auto insurance, but by many insurance companies and agents, and by state insurance regulators as well. No matter how unfair the old system might have been, it was defended, especially by plaintiffs' lawyers who specialize in auto accident cases. The advocates of change ultimately prevailed in a number of states, and we now have "no-fault" auto insurance laws.

No-fault bears some resemblance to the medical payments coverage in a standard auto policy, except that it's more inclusive—and it's compulsory in those states that have adopted it. It enables people to collect from their own insurance companies if they're hurt in auto accidents, no matter who is at fault.

The people who pushed the idea back in the 1960s, before no-fault became a reality—people like law professors Robert Keeton of Harvard and Jeffrey O'Connell of the University of Illinois—envisioned a system that would virtually eliminate the need for lawsuits and lengthy, costly disputes.

But "the right to sue" was an idea that wouldn't die. Many people regarded it almost as a Constitutional right. So the lawmakers who ultimately put no-fault into motion in many states compromised by saying, in effect, "Okay, you can be sued or sue, but only under certain conditions."

Massachusetts was the first state to adopt a no-fault law, which went into effect Jan. 1, 1971 (Puerto Rico had adopted a form of no-fault in 1970). A number of other states followed suit over the next five years.

The laws vary from state to state. A few states have a diluted form of no-fault which places no restrictions on lawsuits. But the states which do impose restrictions all take a hard line on one point: All registered car owners must have what is commonly known as the "personal injury protection" of no-fault insurance.

Generally, if you're the victim, no-fault insurance permits you to recover from your own insurance company such losses

All About No-Fault

as medical and hospital expenses, lost income and other unavoidable expenses growing out of accident-related injuries. It prohibits liability lawsuits except for injuries or medical expenses over a certain degree, such as disfigurement or permanent impairment, or a certain dollar amount (called the "threshold").

Main points on which the laws differ:

- Amount to be paid for medical expenses or for funeral and burial expenses.
- Amount of lost income to be paid to a wage earner.
- Amount to be paid for a person hired to perform necessary services that a homemaker, or other non-income producer, is unable to handle.
- Benefits to be paid to survivors.
- Conditions that allow for lawsuits.

One aspect of no-fault that some people find confusing is the fact that—except in one state—it doesn't apply to property damage. The exception is Michigan, which requires insurers to provide at least a million dollars of coverage, without need to sue, for damage caused by car owners to other people's property as a result of accidents which happen in that state—but not out-of-state. With that exception, damage you cause to other people's property with an automobile can be insured only under property damage liability insurance. Damage to your own car comes under your collision and comprehensive coverages.

Don't make the mistake of one driver who failed to collect information from the other motorist in an accident because he thought each person's no-fault insurance would cover his own damages, regardless of who was at fault. Damage to his car was paid for by his collision insurance but he lost his deductible because his insurer was unable to get the money back from the other driver's insurer.

Fifteen states (and the District of Columbia) have full no-fault laws—that is, laws under which car owners give up their right to sue under certain conditions in exchange for the certainty of knowing that if they are hurt in auto accidents their own insurers will pay their medical and other costs as long as their benefits hold out. Those states are:

Colorado	Kansas	New Jersey
Connecticut	Kentucky	New York
Florida	Massachusetts	North Dakota
Georgia	Michigan	Pennsylvania
Hawaii	Minnesota	Utah

Registered car owners must carry both personal injury protection (no-fault) insurance and liability insurance in all of those states except Florida, where liability insurance is optional. Three other states, Delaware, Maryland and Oregon, require both no-fault and liability insurance, but allow lawsuits without thresholds or other restrictions.

Motorcycle owners, as well as car owners, are required to have no-fault coverage in Delaware and Hawaii.

If you have standard auto liability insurance of your own, you probably won't have to worry about these varying requirements when driving in other states. Most insurers interpret their own liability policies as meeting the minimum requirements of every state. If you are concerned, check with your insurer.

Generally, your no-fault insurance covers you, live-in members of your family, occupants of your car and pedestrians struck by your car, wherever you're driving. You and family members usually are covered if you're hit by somebody else's car.

The restrictions on lawsuits apply to any accident that happens in your own state. If you live in New Jersey, for example, and injure a visiting motorist from Ohio (which doesn't have no-fault), the victim couldn't sue you unless one of the conditions for lawsuits was met. But if the accident happened in Ohio, you would be subject to the laws of that state and wouldn't enjoy the same immunity from a lawsuit as in your own home state.

Who Pays First?

Although the whole idea of no-fault is to permit you to avoid delay and red tape if you're injured in an accident, state laws have given some other coverages precedence, mainly to guard against duplicate payments. Most states provide that if your injury is work-related, the amount of your no-fault payments will be reduced by any amount of workers' compensation payments to which you may be entitled. If you are hurt in a work-related auto accident, be sure to file your workers' compensation claim promptly, without waiting for settlement of your auto claim, as there have been cases where claimants lost out by filing after the deadline, typically 60 days after the injury.

Also given precedence in some cases are such coverages as Social Security, disability benefits, Medicare, Medicaid and other benefits provided under state or federal laws.

Pennsylvania has a unique approach that can save drivers premium dollars. Its law gives motor vehicle owners the option of electing in advance to provide for basic benefits through other accident, health or disability policies. If they choose that route, their no-fault insurance will pay expenses only over and above those provided by the other source and the no-fault premium will be reduced accordingly.

If you live in a no-fault state, you might have the impression you'd be eligible for benefits under both your own policy and that of the other driver. But you won't collect from both. Each state's law spells out an order of precedence.

In a majority of the states which have no-fault insurance, the coverage "follows the car," which means that benefits are paid by the insurer of the vehicle you are occupying or struck by.

In the other states—Arkansas, Connecticut, Florida, Michigan, New Jersey and Pennsylvania—the coverage generally "follows the family." That means that regardless of whether your car is involved or not, your own insurer will handle your claim. If you aren't insured, the other driver's company takes over.

How Lawsuit Restrictions Work

The conditions governing the right to sue after an auto accident differ considerably among the no-fault states. The only constant is death; if you should be killed, your estate or survivors still have the right to sue, even if yours is a no-fault state.

Generally, the idea is to put a lid on lawsuits growing out of relatively minor accidents. Some states say you can't sue at all for your economic losses (the actual dollar damages covered by your policy) unless and until your insurance runs out.

All the no-fault states say you can't sue for non-economic damages, such as pain and suffering, unless certain conditions are met. In every case, the conditions for suit include serious injuries, although there is no uniformity as to what constitutes a serious injury. Depending on the state, it might be anything from a fracture to disfigurement—in some cases, a significant disfigurement, such as the loss of an arm, a leg or eyesight.

A few states stop right there. They have what is referred to as a verbal threshold for suits, which means that victims may sue only if their injuries are of a type described in words in the law. Michigan, for example, permits lawsuits based on injuries only if the injury results in death, permanent serious disfigurement or serious impairment.

A greater number of states add one or both of two other conditions. They leave the way open for lawsuits if the victim:

- Has medical expenses exceeding a stated dollar amount (ranging from a low of $200 to a high of $4,000), or
- Is disabled for a specific period of time (which also varies by state).

The so-called dollar threshold has come under strong attack in many states. Critics argue it's not an effective bar to lawsuits because in these days of soaring hospital and medical costs it's not much of a trick to run up a bill of $200 or $500, or more. And it's easy to see the advantages of crossing the threshold for the injured person, who then can take aim on a sizable settlement by way of the legal process, as well as for the victim's lawyer, who stands to collect about one-third of any award in the form of a contingency fee.

The losers are the other people who carry auto insurance, since the more money insurance companies have to pay out in claims and settlements, the more they have to charge their policyholders.

In general, no-fault laws tend to lower your auto insurance premium by eliminating much costly litigation in auto accident cases. But if you get involved in an accident you will probably consider the biggest benefit to be the elimination of worry and delay in getting your losses paid.

What no-fault can mean in human terms is illustrated by two contrasting cases, one in a no-fault state, Michigan, and the other in a non-no-fault state, New Mexico.

A Michigan girl, 17, was riding with friends in a pickup truck when the driver lost control and hit a tree. She suffered a severed spinal cord which left her permanently paralyzed from the waist down. Her father's no-fault insurance picked up the expenses—more than $50,000 for medical care, rehabilitation and three years of lost wages. In addition, the insurance company provided continuing care which saw her through a two-year college course and placement in a steady clerical job. She was also able to sue the driver and obtain a settlement for her pain and suffering. Meanwhile, her family was spared the financial ordeal of her care. This saved them from going into debt and enabled them to devote their time and energy to helping her adjust to a changed life.

The New Mexico accident was quite similar. It happened to an 18-year-old boy. His spinal cord was injured and he became a paraplegic when the car in which he was a passenger skidded on wet pavement and flipped over. But his situation was far different. His family had no medical insurance, and the driver of the car had no medical payments coverage and only minimum liability insurance. According to the victim's statement, all members of his family—his mother, her three brothers and two sisters, his divorced father and his brother—worked to help pay his bills. Family savings were wiped out. Eventually, the injured man made his way through college and graduate school, and into a steady job. "But over all," he said, "I have detailed a set of circumstances that should not be duplicated."

CONSUMER TIPS

No-Fault Facts

No-fault auto insurance bears some resemblance to the medical payments coverage in a standard auto policy. It differs in two ways:

- ☐ No-fault is more inclusive than medical payments, covering such losses as medical and hospital expenses, lost income and other unavoidable expenses growing out of accident-related injuries.
- ☐ No-fault is compulsory in the states which have adopted no-fault laws.

What no-fault does is to enable you to collect from your own insurance company if you're injured in an auto accident, no matter who is at fault.

No-fault laws vary from state to state. The laws differ mainly around the following points:

- ☐ Amount to be paid for medical expenses or for funeral and burial expenses.
- ☐ Amount of lost income to be paid to a wage earner.
- ☐ Amount to be paid for a person hired to perform necessary services that a homemaker, or other non-income producer, is unable to handle.
- ☐ Benefits to be paid to survivors.
- ☐ Conditions that allow for lawsuits.

The original idea behind no-fault was the elimination of lengthy and costly lawsuits for accident victims. In practice, some no-fault states allow lawsuits, but only under certain conditions. The only constant is death; if you are killed, your survivors still have the right to sue, even if you live in a no-fault state. Your agent or company or your state insurance department can provide you with a brochure setting out the details of your state's law.

Consumers often make the mistake of thinking no-fault coverage includes property damage. That is true only in the state of Michigan. If you live in any of the other 49 states, you should be aware that no-fault applies only to personal injury. Don't depend on no-fault to replace your collision or comprehensive coverage.

No-fault sometimes pays only after other coverages have been exhausted. State laws have given other coverages **precedence**, mainly to guard against duplicate payments.

Coverages taking **precedence** include:

- ☐ Workers' Compensation
- ☐ Social Security
- ☐ Disability Benefits
- ☐ Medicare
- ☐ Medicaid
- ☐ Other benefits provided under state or federal laws

No-fault laws eliminate much costly litigation in auto accident cases. In addition to this economic benefit, no-fault helps to eliminate needless worry and delay in getting your losses paid.

Claims under no-fault insurance should be filed as soon as possible after an accident, just as should all insurance claims.

CHAPTER 12

HOW TO HOLD YOUR AUTO PREMIUMS DOWN

Like most people, you probably fume when you pay your auto insurance premium. It's hard to see what you're getting for your money. And in any event, it seems like too much. Why is your premium so high? And why is your premium different from other people's? You probably could poll a dozen of your friends and find that no two of them pay the same amount for auto insurance. Why do you pay more than your next-door neighbor? Or less than the father of the teen-ager down the block?

An insurance agent compared his customers' opinions about the cost of auto insurance with their attitudes toward the cost of groceries:

"People only grumble about the rising cost of groceries," he said. "They become used to it, and at least they're getting something to eat. But car insurance is a big ticket item that a lot of people don't think much about until they get their bill. Then they see red. It's the biggest source of complaint among my customers."

"All I can do," the agent adds sadly, "is explain that auto insurance is hit by inflation like everything else—only harder."

Not only has the cost gone up just to maintain the same coverage, but inflation has left many policyholders seriously underinsured. They should have been increasing the amounts of their coverage to keep up with the cost of living.

If you could thumb through the claims records of almost any auto insurer, you'd find cases much like these, paraphrased from the files of a leading company:

- Claimant: Housewife, 43, mother of two. Right leg fractured and both kneecaps partially severed when her husband lost control of the car on a bend in the road early on a January evening (weather dry and clear) and crashed head-on into a tree. Husband, who investigators said had been drinking and was driving at an excessive speed, was killed. After several surgical operations and reconstruction of both kneecaps, the woman still needed a walker to move about, plus physical therapy treatments twice a week. Claim closed after payment of full liability limit of $20,000, including an advancement of $5,000 to cover early expenses. Anticipated total cost of medical treatments: $50,000—or $30,000 more than the available amount of coverage. (In some states where interspousal suits are barred, the wife would not have been eligible for any recovery.)
- Claimant: (Survivor of) engineer, male, 27, with three dependents. Killed shortly after 6 p.m. on a clear July evening when his car was hit broadside by another vehicle whose driver (the insured) failed to stop for a stop sign. "The claimant might have assumed some liability due to the speed of the vehicles at the time of impact." Injury value estimated at far more than the $50,000 liability insurance applicable. At writing, claimant had refused to settle for policy limit and case was in litigation.

The effect of such cases is that either the underinsured policyholders or the claimants are left to deal with major financial burdens on their own.

The Effect of Inflation

The Consumer Price Index (CPI), published by the U.S. Bureau of Labor Statistics, is generally accepted as the best available yardstick of living costs. Each month, it compares the current average costs of consumer goods and services with the costs of the same items during a base period in the past.

As of December 1982, the average cost of auto insurance had increased 188 percent since 1967, while the overall cost of living, which takes just about everything into account, had climbed by 192 percent. Auto premiums were up because inflation drove up the costs of items for which auto insurance pays. For example:

Medical care (all aspects)—up 244 percent.

Physicians' services—up 236 percent.

Hospital rooms—up 488 percent.

Auto repairs and maintenance—up 223 percent.

Even the cost of the proverbial fender-bender accident comes high today. If you can get by with ironing out a wrinkle or hammering a dent back into shape, fine. But the cost of replacing a whole fender assembly can equal your entire auto insurance premium, or a good chunk of it. The cost of a fender for one medium-sized car late in 1982 came to $399, including labor.

Don't forget, inflation or not, it costs a lot more to put together an entire car in the repair shop than on the assembly line. It would cost more than $26,000 to replace all the parts of a popular subcompact that you could buy new for about $7,000 in 1982.

Yet, even though the cost of auto insurance is soaring, there are things you can do to hold your premiums down.

How Your Auto Premium is Figured

Actually, what you pay to insure your car depends on a number of variables. Where you live. The kind of car you drive. How much you drive it and for what purpose. Who will be driving your car. What kind of driving records they have. And yes, your choice of an insurance company.

Calculating auto insurance rates is a complicated process. Under state regulatory rules, an insurance company's premium collections are supposed to take care not only of the cost of adjusting and paying claims but of the company's administrative, marketing and other overhead expenses as well.

Insurers can predict pretty accurately what their overhead costs are going to be. Forecasting claims costs is the tough part. For this, the companies or the statistical organizations to which they subscribe hire mathematical experts called actuaries.

Since the accidents, thefts and other losses that premiums primarily are intended to pay for haven't happened yet, the actuaries who figure the rates have to come up with projections of:

1. How many accidents and other types of losses insured drivers will have during the period the rates will be in effect, and
2. How much those losses will cost.

Rates have to be determined for each of the auto insurance coverages, and further broken down to reflect the experience of various areas in each state and also the loss records of the different classifications of drivers.

Let's take a closer look at some of these factors.

Territories

A short drive from the bustling city of Albany, N.Y., will place you in the foothills of the Adirondacks in one direction, the northern slopes of the Catskills in another. Either way, traffic will have thinned out appreciably.

If you live in one of these outlying areas, therefore, your exposure to accidents will be less than if you live in the city. And the bills for having yourself or your car patched up, if you do have a crack-up, will probably be less. For these reasons, auto insurers will probably charge you a lower premium.

Because of such differences in risk, auto insurers divide each state into rating territories. A rating territory may be a town, a large city or part of a city, a county or some other geographical subdivision. Boundaries are drawn so that population density, traffic conditions and various accident-related factors are more or less uniform within each territory.

While the record of claims is a major factor in the ratemaking process, a three-car pileup right in front of your house won't necessarily be "charged" against your community in the insurance rates. For example, if you live in Omaha and the driver who caused the accident hails from Chicago, the claim will be charged to Chicago. Insured losses are charged back to the home bases of at-fault drivers, and this one ultimately would affect the rates in Chicago.

Nevertheless, most local accidents will be charged against your community because most driving is done close to home. In fact, more than half of all auto trips cover less than five miles. The number of claims in your community is affected both by traffic and road conditions and by how strictly your traffic laws are enforced.

The other big factor that affects any given community's insurance rates is the average cost of insurance claims in that particular community. Car repair charges, hospital and medical bills and jury awards in personal injury cases—all factors in the cost of auto insurance—vary widely from one area to another. So do personal attitudes about filing insurance claims that are honest and reasonable, another factor that influences ultimate costs.

Driver Classification

The insurance costs of a community are shared by all the insured drivers who live there. But they're not shared equally, as most people are keenly aware.

There's no reason why insurance companies couldn't charge the same price for everybody. The elderly couple who use their car to go grocery shopping, get to church on Sunday and take an occasional spin in the country. The young hotrodder who delights in zipping along country roads at 75 miles an hour. The confirmed alcoholic who smashes up a couple of cars each year.

Although there are some people who think such a flat rate would be fair, it really wouldn't make much sense. It would mean the elderly couple would have to pay more, although they represent a relatively low accident risk, so the other, high-risk drivers could pay less than their cost to the system. When you look at it that way, a flat-rate premium definitely would be unfair to the safer drivers. And that's the way the insurance companies see it.

Thus, car owners and their families are grouped into a number of classifications reflecting their statistical likelihood of having accidents or other insured losses. Where you fit in, and the size of your premium, may depend on several personal factors that have little to do with where you live.

Age—Most parents have apoplexy when they get their first auto insurance bill after their teen-ager acquires his or her driver's license. Usually, there's a big difference in the premium—and the difference is way up. Should the teen-ager want his own car, on his own policy, the premium would probably be one few teen-agers could afford.

Not that teen-agers necessarily are bad drivers. Most of them have excellent reflexes—as they will be quick to tell you—and they can handle a car as though it were a physical extension of themselves—which in a sense it is. Unfortunately, when you take the entire age group 17 and under through 24, numbering some 32 million, the statistics show that young people have a lot more than their share of accidents.

Whether you chalk it up to inexperience, recklessness, immature judgment or—especially among youths in their late teens—drinking, the accident rate among young drivers is disproportionately high. National Safety Council data show that in 1981 (and the statistics don't change much from year to year):

- Drivers under 25 made up 21.7 percent of the nation's motoring public, but accounted for 36 percent of all drivers involved in accidents.
- The accident involvement rate of that group was more than double that of drivers 35 and older.
- Auto accidents were by far the leading cause of death among men and women 15 through 24 years old, claiming nearly as many lives among people in that age bracket as all other causes combined.

Drinking is a factor in at least half of all motor vehicle accidents and its particular effect on young people is reflected dramatically by the experience of one state which lowered and then raised its minimum legal drinking age. After Michigan lowered its legal drinking age from 21 to 18 in 1972, a study covering an eight-year period (four years before and four years after the change) showed a 35 percent increase in alcohol-related accidents involving drivers 18 to 21 years old. In 1978, Michigan restored the former minimum drinking age of 21, and in the following year the number of 18- to 21-year-old drivers involved in alcohol related accidents dropped by 31 percent.

It's because young drivers as a group (without singling out any individuals) cause more than their share of accidents that higher premiums usually are assigned to cars which are available to them on a regular basis. If it seems unfair to penalize a given teen-ager whose driving record is still clean (perhaps because he has been driving a very short time), he must blame his peers in his age group—whose collective record is none too good.

Although auto insurance for a teen-ager usually is going to cost extra, there still are ways you can save money. It's cheaper to carry the youngster as an additional driver on a parent's policy than to insure him or her separately. And the cost can be lowered further if you can limit the youngster's driving to less than 10 percent (or some other minor percentage depending on the company) of the total use the family car gets. That way, he or she can be listed as an occasional, rather than principal, driver.

There are other factors. If a youngster is more than 100 miles away at school most of the time, that's often good for a discount. If a young driver is in the military service, the family's premium probably won't be affected at all unless he or she gets home often enough to make frequent use of the car.

Incidentally, while you're paying more now, you can look forward to a break later. Age also has a bearing on insurance premiums further on in life. Many companies give a premium credit to drivers 65 and older, who generally do less driving

Rates Go Down As Young Drivers Grow Older*

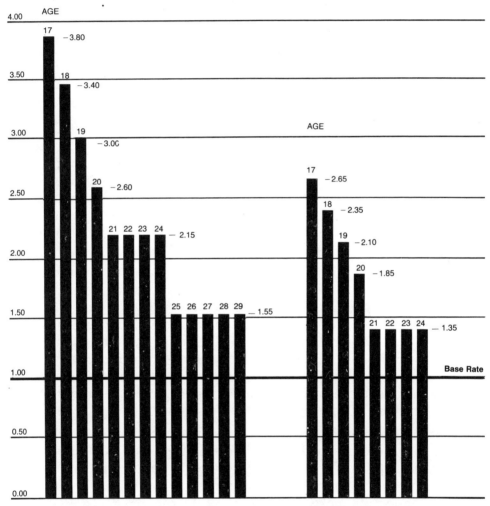

Unmarried male operator under age 30 who owns or is a principal operator of the automobile

Unmarried male operator under age 25 who is not the owner nor principal operator

*Based on a driver classification plan used by a large segment of the business in many states. These comparisons of gradations are for private passenger cars used for pleasure where all operators have clean driving records. Adjustment in premiums are made for cars used to drive to work, used for business, or used on a farm. Adjustments are to make the younger operators with driver training credit, drivers with "unclean" driving records, and owners of more than one car. In many of the premium discounts are available to students with outstanding scholastic records and to unmarried students attending school 100 or miles from home.

Rates Go Down As Young Drivers Grow Older*

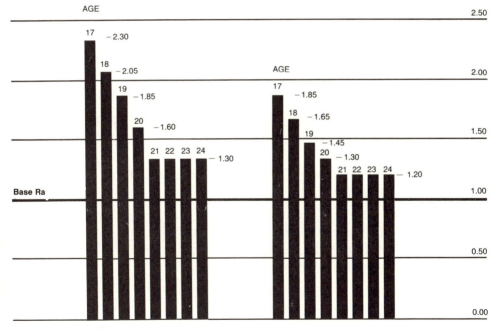

*Based on a driver classification plan used by a large segment of the business in many states. These comparisons of gradations are for private passenger cars used for pleasure where all operators have clean driving records. Adjustment in premiums are made for cars used to drive to work, used for business, or used on a farm. Adjustments are to make the younger operators with driver training credit, drivers with "unclean" driving records, and owners of more than one car. In many of the premium discounts are available to students with outstanding scholastic records and to unmarried students attending school 100 or miles from home.

than their juniors and are involved on the average in fewer accidents.

Sex and Marital Status—In the under-25 age group, women and married men have better driving records, collectively, than unmarried males. So most companies give them a relative price break on their car insurance.

Over the years, women have had a lower overall accident rate than men, although the gap has been narrowing. When it comes to fatal accidents, men have a much worse record. On the basis of miles driven, men are involved as drivers in about twice as many fatal accidents as women.

Thus, under certain conditions, women can continue to enjoy a price break through their adult years. Many companies give a lower rate for women 30 to 64 years old who either live alone or are the only drivers in their households.

Marriage seems to bring out the best in young drivers, if insurance records are a good indicator. Possibly people drive more carefully when they have someone else to think about. As a result, many companies charge young men less if they're married, and treat married young women as though they'd already moved out of the high-risk "youthful driver" bracket.

Premium differentials for age, sex and marital status of young drivers vary from one company to another. The cost often is scaled down as a young driver grows older and becomes more experienced in driving.

With most young drivers, 25 is the "magic age" at which they no longer are considered to be youthful drivers—which means they qualify for a substantial reduction in their auto insurance premiums when they turn 25.

Nevertheless, many companies extend that cut-off age to 30 for unmarried men who own or are the principal drivers of cars. Statistics indicate that they take longer than other young drivers to shed their accident-proneness.

A few states, by law, forbid auto insurers to apply price differentials to car owners on the basis of their age, sex and martial status. One effect is that young single male drivers in those states pay less for their auto insurance, but that forces other car owners to pay more.

Education—Young drivers often can save dollars on their premiums in one or more of three ways:

- By completing an approved driver education course;
- By maintaining a high scholastic average;
- By attending a school 100 or more miles from home.

Insurance companies have been supporters of driver education for many years, and many companies back up their faith by offering discounts, obtainable by presenting a certificate of completion of an approved course. Discounts vary by company, but typically are about 10 percent, and usually are offered through age 20.

Insurers also have found a correlation between a good scholastic standing and safe driving (whether because good students drive more responsibly, or just drive less). Discounts are available to high school or college students under 25 (usually they must be at least 16 or high school juniors to qualify) as long as they meet these requirements:

- Rank in the top fifth of the class;
- Maintain at least a "B" average (or a "3" average in a school using a 4-point grading system); or
- Are on the dean's list or honor roll.

Driver education and good student discounts are not mutually exclusive. A student with driver training who also has high grades may qualify for both discounts.

The third education-related credit is offered by many companies to unmarried students who attend and live at a school more than 100 road miles from home, or from wherever the family car is kept. The theory is that the student will be using the car only occasionally rather than on a regular basis.

Individual Safety Records—Here is where a driver has significant control over his or her insurance bill. Or, to put it more accurately, this is where members of a car owner's family have influence over what the owner pays to insure the vehicle. If a family has a clean driving record for a given number of years some companies will offer a reward, either in the form of a premium discount or by "forgiving" the first accident. But generally, if you, or someone in your family, have a "chargeable" accident or are convicted of a serious driving violation, you can expect an increase in your insurance premium the next time around. The premise is that people who have had accidents or serious violations are more likely to have accidents in the future than are drivers with clean records. Usually this surcharge remains in effect for about three years and then is discontinued if there hasn't been another accident or violation during that period.

As with other pricing procedures, rating on the basis of personal driving records varies by company and by state. Usually, points are assigned for each chargeable accident or violation, and the amount of any premium increase is pegged to the number of points family members have accrued.

Accidents usually aren't counted against you if you're using something other than a family-type vehicle—such as a motorcycle or a bulldozer. But a conviction for drunk driving, speeding or some other serious offense will be charged against you no matter what you are driving when arrested.

Typically, here's how that point system works as spelled out in one company's policy attachment.

For violations resulting in convictions:

- 3 points for drunk driving or driving while under the influence of drugs, hit-and-run driving, motor vehicle homicide or assault, or driving while license is suspended or revoked.
- 2 points for the accumulation of points under a state point system or a series of convictions.
- 1 point for two or more chargeable property damage accidents for which points haven't already been assigned.

For accidents:

- 1 point for an accident resulting in bodily injury, death or more than a certain amount of property damage.
- 1 point for two or more chargeable property damage accidents for which points haven't already been assigned.

An additional point is charged if the principal operator of the car has been licensed for less than three years, unless the operator's premium is already subject to a surcharge because of an accident.

Not all companies use the same formula, but those that do typically charge an additional 40 percent of your premium for one point, 90 percent for two, 150 percent for three and 220 percent for four.

What's a chargeable accident? Insurance companies answer that question by setting forth a list of situations under which accidents are not chargeable and therefore don't affect the cost of your coverage. Generally, an accident won't be counted against you if you can demonstrate—following the guidelines—that it wasn't your fault.

One situation in which a crash won't affect your premium is an accident that happens while you're responding to an emergency call as a paid or volunteer member of a police or fire department, first-aid squad or other law enforcement agency. But you won't get the same consideration if you wreck the car on the way home.

Going back to that same (typical) policy, an accident also will be excused if:

- Your car was lawfully parked;
- You're reimbursed by, or obtain a judgment against, another person responsible for the accident;
- Your car is struck in the rear by another vehicle and nobody in your household is convicted of a moving traffic violation as a result;
- The other driver is convicted of a moving traffic violation and the family member who was driving your car isn't;
- Your car is damaged by a hit-and-run driver (but the accident has to be reported to proper authorities within 24 hours);
- The accident involves damage by contact with animals or fowl, or was caused by flying gravel, missiles or falling objects;
- The accident results in a claim or payment under personal injury protection (no-fault) coverage and doesn't otherwise qualify for the assignment of a point. (This kind of stipulation wouldn't apply, of course, in a state which doesn't have a no-fault law.)

An important point you might want to keep in mind is that these premium surcharges are applied only because your car regularly is used by or available to a family member who has had a chargeable highway misadventure. Therefore, your premium is subject to adjustment, downward or upward, if:

- Someone who has had an accident or a chargeable violation is added to the household, or
- The person who had the accident or violation leaves the household.

You should let your agent or company know as promptly as possible if your family's overall driving record is affected by such things as a divorce, a child getting married and moving away from home, or the acquisition of a new spouse.

Use of Car—Cars that are used for business, even if only to drive to and from work, usually have a greater exposure to accidents than cars used only for pleasure. That's because usually and regularly they are driven at times and in places where traffic is heavier—often in congested rush-hour traffic.

Insurance companies make several distinctions, reflected in the premium, based on how a car is used.

Generally, you'll pay a little more if you drive your car to and from work a one-way distance of three to 15 miles. More yet if you drive to and from work a one-way distance of 15 or more miles or if you use your car for business purposes throughout the day. The differentials are applied even if you drive only part way to work, such as to a railroad station or a bus terminal.

If you're looking for a way to get a break on your premium, consider joining or forming a car pool. For instance, many companies will charge you less if you do the driving no more than two days a week or two out of every five weeks. If such an arrangement is practicable and you can work it out, be sure to let your insurance representative know.

Special concessions usually are made for:

- Farmers, who may qualify for a reduced premium if they don't use their cars for any business other than farming, and

- Clergymen, who are spared any additional "use of car" charge even though they may use their cars for "business" purposes all day.

Some insurance companies differentiate, even among vehicles used strictly for pleasure, on the basis of the estimated number of miles driven over the course of a year. The rationale is that, on average, cars which are driven for greater distances tend to have a higher frequency of accidents.

Type of Car—You can save some money depending on the type car you drive. Some cars cost more to repair than others. The most expensive models, for example, generally are more expensive to fix, or to replace, than lower-price models. That is taken into account in the portions of your premium (for the collision and comprehensive coverages) which pay for damage to, or the loss of, your car. If your car was made in 1980 or later, some states allow rates to be based directly on the overall loss experience of that model. You pay a lower premium if your model has a record of lower repair costs from accidents.

Number of Cars in the Family—When a family has two or more cars, each usually is driven less than a single car owned by a one-car family. If you have two or more cars and insure them under the same policy, you'll pay less than if you insure each individually with a different company.

More Incentives for Safety—Many companies offer premium credits, or discounts, in recognition of special efforts on your part to improve your odds against auto-related insurance claims.

A number of insurers will give you a price break on the medical payment or personal injury protection (no-fault) portion of your premium if you buy a car equipped with "passive" restraints. These may be air bags (available on some European luxury cars) designed to inflate instantly after a front-end crash to prevent front-seat occupants from being thrown against the steering wheel, dashboard or windshield. Or, more likely, they may be seat belts that wrap around you automatically when you close the door.

Installation of special equipment designed to foil would-be car thieves can gain you a discount in a few states on the portion of your premium that pays for comprehensive coverage. To qualify for the discount, your car must be equipped with an interior hood lock plus an alarm and/or a device that makes the fuel, ignition or starting system inoperable.

You may be eligible for a premium discount if you've completed a certified course in defensive driving. While the provisions may vary somewhat, the laws of six states—Arkansas, Delaware, Illinois, Louisiana, New York and Texas—require that such a discount be offered. The Illinois law, which became effective in mid-1982, calls for a discount, good for three years, to drivers over age 55 who complete the widely used defensive driving course of the National Safety Council.

A few companies will even cut the price of car insurance for clients who don't smoke, presumably on the grounds that they're less likely to become distracted while driving. Whatever the reason, an executive of one such company said the collective insurance claims record of his firm's non-smoking policyholders has been good enough to justify the discount. While the company does a little spot-checking, the discount usually is offered on faith.

A few insurers write auto insurance exclusively for non-drinkers at a reduced premium. If you abstain you might want to save money this way. The standard policy is offered to car owners who attest on their applications that they don't, and don't plan to, imbibe except during a ritual that is part of a religious service in a church or temple.

The Bottom Line on Costs

Precisely how much will all these discounts and surcharges affect the price of your auto insurance? How much more will your policy cost when your son turns 17 and gets his driver's license? How much of a premium credit will you get if your child is on the honor roll at school? How many more dollars will an accident or a violation point cost you? There are no hard and fast answers. Different companies handle things differently.

The table on Page 142 will give you a idea of how the various discounts work, as reflected by a survey of 10 major auto insurance companies, most of which operate countrywide.

Bear in mind that when it comes to price adjustments for young drivers and especially those who are still living at home, it's usually the parents' premium that's affected. If you're the parent, you might insist that young Johnny or Mary scrape up the money for the difference in cost, but it's your policy and you're responsible for paying the premium.

The various credits and debits are all factored in separately as multipliers of "base rates" for the coverages to which they apply. The base rate for your territory is on file with your state insurance department.

It's not as complex as it sounds. Imagine a family with a standard-performance car which is used for pleasure only. They live in a territory where the base rates for all of their coverages total $200, but their 17-year-old son Tommy got his driver's license shortly before their policy was last renewed. Because of Tommy, the premium is now computed at 2.65 times the base rate, or $530.

Now Dad, on his way home from a late-night poker game, dozes off at the wheel, swerves into a tree and winds up with a broken collarbone, As a result, his premium becomes subject to a 40 percent surcharge the next time around. But the 40 percent is applied to the $200 base rate, not to the existing premium of $530.

New premium: $610 ($530 plus 40 percent of $200, or $80).

Any credits Tommy might earn, say by completing a driver education course or by maintaining high grades at school, also would be applied to the base rate, not to the total premium.

What's a Discount Worth?

The size and application of special premium credits offered by car insurers vary. They're not all available from all insurers in all states. If a driver in your family qualifies for one or more discounts, be sure to check with your own insurance representative about your eligibility for a premium reduction.

Whatever else you do about your auto insurance costs (such as shopping around for the best rates), do inquire from your agent or company as to what discounts are available. Review your eligibility for any or all of them. You may be eligible for more than one.

Do the same for other drivers in your family. For example, you can't expect the insurance company to keep up with your

Discounts Offered By 10 Leading Auto Insurers

Type of Discount	Companies (of the 10)	Savings Offered	Coverages Applicable to
Driver Training	9	Range 5–40%; most commonly 10%. Offered by one company only in Nebraska and Pennsylvania	Generally, all coverages (may not apply to UM)
Good Student	9	Range 5–30%; commonly 25%	Generally, all coverages (may not apply to UM)
Student Away at School	10	Some companies put policyholders in a lower price bracket; others offer percentage discounts, usually 10–15%.	Most coverages
Car Pool	10	Some companies put policyholders in a lower price bracket others offer percentage discounts of 15–20%.	Usually, all coverages
Multi-Car	10	Range 10–25%; usually about 15%	Generally, all coverages (may not apply to UM).
Passive Restraints	7	Range 10–30%	Commonly applies to no-fault and medical payments only.
Anti-Theft Devices	9	Range 5–15%. Most companies offer only in a few states including Illinois, Massachusetts and New York.	Usually, comprehensive only.
Female Age 30–64, Sole Driver in Household	4	Usually 10%	All coverages
Senior Citizen	9	Range 5–15%. Usually offered at age 65; by one company at age 55.	Most coverages
Farmer	10	Range 10–30%; most commonly 10%	Generally, all coverages (may not apply to UM)
Defensive Driving	7	Range 10–15%. Usually offered in only a few states, including Arkansas, Delaware, Illinois, Louisiana, New York and Texas.	Most coverages (with some companies, liability only).
Non-smoker	1	12–25%	Liability, collision, no-fault, medical payments

son's progress in school unless you report that information. If your son is going to college more than 100 miles from home, report that fact. If he's on the dean's list, say so. Your agent or a company representative will take the information. A phone call is easiest; put it in writing if you're asked. The saving is well worth the few minutes it takes to make the call.

You also can do some things in the public arena to hold your car insurance cost down. Back legislation to improve vehicle and highway safety. Lend your support to the public campaign to make drunk driving socially unacceptable. And, by all means, obey the law yourself. Fewer accidents will mean lower auto insurance premiums for you.

CONSUMER TIPS

Keep Your Costs at a Minimum

To do what you can to hold your auto insurance premiums down, you'll want to be certain to take advantage of **all available discounts**.

Check the discounts below for which you may be eligible. Ask your agent or company representative about each item you check:

- ☐ Driver Training
- ☐ Good Student
- ☐ Student Away at College
- ☐ Car Pool
- ☐ Multi-Car
- ☐ Passive Restraints
- ☐ Anti-Theft Devices
- ☐ Female Age 30-64, Sole Driver in Household
- ☐ Senior Citizen
- ☐ Farmer
- ☐ Defensive Driving
- ☐ Non-Smoker
- ☐ Non-Drinker

Save dollars by avoiding traffic violations and insist that other drivers in your family do the same. (Remember that a conviction for drunk driving, speeding or ignoring a stop sign means points charged against your driving record—and an avoidable addition to your premium.)

Insure your teenager as an additional driver on your own policy rather than separately.

Lower costs further by limiting your youngsters' driving to a small percentage of the total use of the family car.

Take action, by informing your agent or company, to have any accident excused from your record if:

- ☐ Your car was lawfully parked.
- ☐ You're reimbursed by, or obtain a judgment against, another person responsible for the accident.
- ☐ Your car is struck in the rear by another vehicle and nobody in your household is convicted of a moving traffic violation as a result.

- ☐ The other driver is convicted of a moving violation and the family member who was driving your car isn't.
- ☐ Your car is damaged by a hit-and-run driver (but the accident has to be reported to proper authorities within 24 hours)
- ☐ The accident involves damage by contact with animals or fowl, or was caused by flying gravel, missiles or falling objects.
- ☐ The accident results in a claim or payment under personal injury protection (no-fault) coverage and doesn't otherwise qualify for the assignment of a point.

Let your agent or company know of any changes in your family or lifestyle which could affect your premium. Examples: a divorce, marriage, a child's departure for college, a change in the way you use your car.

Make insurance costs a matter of importance when you are considering a new car. You may want that conservative sedan rather than the flashy sports racer after you've evaluated what impact it could have on your insurance premiums.

CHAPTER 13

WHAT IF THE COMPANY TURNS YOU DOWN?

Anyone can get auto insurance. That doesn't necessarily mean you'll get it from the first insurance company you apply to. After all, you can't necessarily count on getting a loan from the first bank you apply to. Before handing out loan money, a bank will size you up as a credit risk. If you don't have a credit record, or if it's not too clean, the bankers may turn you down. Or, they may turn you down because money is tight, because they would rather make business loans than personal loans, because they're "loaned out" at the time, or for other reasons. But just because one bank turns you down, that doesn't mean the bank across the street won't welcome your business.

In a general way, the same thing is true of insurance companies. When you apply for auto insurance, you are sized up as a driving risk. If you have a driving record with some accidents or violations on it, or if you simply haven't been driving or been insured before, the company may turn you down. After all, you're asking them to assume a major financial risk.

What do you do then? Go across the street to another company. Or, if you're dealing with an agent representing several companies, none of which will take you, try another agent.

Even if you draw a total blank in the "voluntary" insurance market, don't despair. You can still get auto insurance, though you may have to pay more for it. Where? Either in the "non-standard" market, or in the "shared" market.

The non-standard market is made up of companies that specialize in hard-to-place driving risks. Usually these companies charge higher premiums than regular insurers. Any agent either should be able to handle your application for this market, or steer you to an available source.

Typical policyholders with non-standard companies, in addition to those with poor driving records, include drivers of high-powered "performance" cars or expensive sports cars, persons considered "high risk" for other reasons, and those who may want more coverage than is provided by the shared market.

The "Shared" Market

A state-by-state network, the "shared" market is what used to be called "assigned risk plans." That's where insurance is provided for people who have trouble getting auto insurance from any one company. It's subsidized by the insurance companies and by their other policyholders. It's subsidized to such an extent, in fact, that in 1981 alone the shared auto insurance market ran a deficit of just over $1 billion—a loss which was made up by the companies doing business in each state—and ultimately by their regular policyholders through higher premiums.

The shared market came into existence as a solution to a practical problem. There's no law, of course, that says everybody who wants a loan from a bank can get one. But there are laws in nearly every state compelling car owners either to have auto insurance or to be able to show financial responsibility. Since most people need their cars—can't live without them, really, in today's world—it is imperative they be able to obtain auto liability insurance.

New Hampshire started the first assigned risk plan in 1938. The idea mushroomed, and through the insurance industry's cooperation every state has had a shared-market auto insurance set-up of one kind or another since 1959.

Through this assuming of risks by the insurers, any licensed driver can get at least enough liability insurance to meet the minimum requirements of state laws. In most states, you can get greater amounts of liability insurance, as well as other coverages such as collision and comprehensive insurance to cover damages to your own car.

Some people think there's a stigma attached to being in an "assigned risk plan." Not really. All kinds of people end up there for all kinds of reasons.

For example, the main thing about Jack Wilkens that set him apart from a lot of other members of the fraternity of "assigned risks" was that he was a long-time employee of the insurance company that canceled—or, technically, declined to renew—his auto insurance policy. Computers know no favorites. Jack's affiliation with the company and his reputation as a "nice guy" weren't enough to wash out the record: a "totaled" car in a single-vehicle accident a few blocks from his home, followed by a conviction for driving while under the influence of a few drinks too many.

Phil Blaine's insurer took it as a matter of course when he swerved off an interstate highway and wrapped his car around a signpost. The decision not to renew his policy came not because of the crash alone, but because Phil then chose to replace his conservative model automobile with a high-powered sports car.

Both Jack and Phil wound up getting their next insurance policies through the shared market. Both made mistakes that made them look like poor driving risks to their insurance companies. The companies exercised the right to turn down their business.

If you buy your insurance through the shared market:
- In most states, the cost is higher than most companies charge for equivalent coverage sold to you voluntarily. On the other hand, in some states the price generally is competitive. In rare instances, you could wind up insured as an "assigned risk" by a company that turned you down earlier.

- The shared market is a little stiffer about applying premium surcharges tacked on because of traffic violations. In New York, for example, some companies insuring voluntarily won't charge you anything extra as the result of a conviction for exceeding the speed limit by less than 15 miles an hour, or for such infractions as running a red light or ignoring a stop sign. But if you're insured through the shared market automobile insurance plan, you'll face the certainty of a premium hike which in some areas can exceed $100 a year.

Getting into the shared market is no problem. Any licensed agent or broker can handle an application.

There are four kinds of shared market systems for auto insurance:

Automobile Insurance Plans—These plans, which are the most commonly used, operate on an assignment system in which each insurer's participation is in proportion to the amount of business it handles voluntarily in the state. The applicant doesn't select the insurer but is assigned to the company that comes next in an order of rotation.

Reinsurance Facilities—In the few states that take this approach, every company is required to accept any applicant for insurance, no matter how bad the applicant's record. But the company can then "reinsure" its poor risks by transferring them to the facility, a paper transaction that has no effect on the way the company handles and services the policy. At the end of the year, each company shares in the total losses of the facility through assessments based on the percentage of the total business written voluntarily in the state. States with reinsurance facilities are Massachusetts, New Hampshire, North Carolina and South Carolina.

Joint Underwriters' Associations—Florida, Hawaii, Michigan and Missouri have JUAs, in which several companies which agree to act as "servicing carriers" write and service all the shared market business. Separate records are kept on this business, and all auto insurers share in any losses, as well as expenses.

State Fund—Maryland, alone among the states, handles its shared market business through its own insurance company.

The Maryland Automobile Insurance Fund (MAIF), which was created by the legislature in 1973, accepts applicants for insurance only after they have been rejected by at least two insurance companies. Private insurers are required by law to subsidize any losses incurred by the fund, and the cost is charged back to their own policyholders.

Countrywide, in any given year, six to eight of every 100 private passenger cars are insured through these various methods. But that's only an average. The fact is that the plans are big operations in only a handful of states, and little more than an accommodation for relatively few car owners in most states.

Overall, the shared market is like a sponge filled with red ink for the auto insurance companies. In a typical recent year, the companies were hit by $134.70 in costs for every $100 in premiums taken in through the shared market. New Jersey and Massachusetts, where about one of every three policyholders is insured through the shared market, accounted for almost 70 percent of the entire $1.04 billion loss.

Critics have charged that the various auto insurance plans are overpopulated with "clean" risks. That insurers use the plans as a convenient parking place for such groups as young drivers, retirees and the unemployed. That the plans—and the rates—are tilted against people who live in the big cities.

To see what the real story was, the insurance industry at the outset of the 1980s commissioned an independent demographic study of the plans' operations in five states: California, Florida, Kansas, Pennsylvania and Virginia. The study showed that 55 percent of the assignees had unsatisfactory driving records, and that another 36 percent were there for one of two other reasons:

1. They had less than three years of driving experience, or
2. They had never carried auto insurance before.

The companies' rationale for turning down drivers in the regular market because of inexperience stems from the fact that newer drivers, regardless of age bracket, represent a greater risk than experienced motorists. Quite simply, they average more accidents. Even when they obtain insurance through regular channels, inexperienced drivers frequently pay a higher premium.

That helps to explain another finding, that drivers under age 25 make up a significant segment of the plans' population. Either because of their inexperience, or accidents and violations already on their records, young drivers make up a far greater percentage of the drivers in the plans than the proportion of all under-25 drivers to the motoring population as a whole. Nevertheless, there are still fewer youthful drivers in the shared market than in the voluntary market. At first glance, it might seem unfair to turn someone down just because he or she hadn't been insured before. But there is a good reason. One shared market executive sums it up this way: "Auto insurers sometimes take an applicant's insurance record into account for much the same reason that a lender wants to have a look at the previous credit history of an applicant for a sizable loan. It's difficult to check out the past driving record of an applicant who has never had insurance."

The study of shared markets showed it is a myth that the plans are largely repositories for unwanted motorists who live in urban areas. Overall, the geographical makeup of the plans' population was about the same as that of the rest of the population. Nevertheless, it's not hard to make a case for turning down an applicant who "garages" his expensive car on a congested street of a city with a high auto theft rate.

The Way Out

Being in the shared market isn't a hopeless fate. An "assignment" to the shared market usually is for a period of three years. If you've kept your record clean during that period, it shouldn't be difficult in most states to find an insurer that will accept you voluntarily and willingly.

Even if you've had an accident or a couple of speeding tickets during the three years, don't hesitate to shop around. The worst that can happen is that you'll be turned down again, in which case you may be assigned to another company through the shared market.

It's possible, especially if your driving record was spotless in the first place, that with a little more perseverance you could have avoided having to obtain your coverage through the shared market. And there's no reason that you have to give up looking even after you've been assigned to an insurer through that market—although you'll be penalized for administrative costs on your refund if you cancel your present policy before it expires. If you feel it's economically feasible, make a change if you can find an insurer that will accept you voluntarily. So, once again, talk to your agent and see what's available.

CONSUMER TIPS

Don't Be Discouraged

- ☐ Don't decide to drive without insurance because a company turned down your application. That's the worst decision you could make.
- ☐ Do go to another company or to several other companies and shop for insurance.
- ☐ Don't hesitate to ask your agent about companies which specialize in hard-to-place driving risks, but know that these companies charge higher premiums than regular insurers.
- ☐ Do recognize that you, as a licensed driver, can get at least enough liability insurance to meet state requirements through the system known as the shared market. But be aware that:
  - ☐ The cost is higher in most states.
  - ☐ You will be surcharged for traffic violations.
- ☐ Don't stop shopping for insurance once you're placed in the shared market. You might be fortunate enough to find a bargain!
- ☐ Do drive carefully. A clean driving record for three years will mean you're likely to be able to find an insurance company and save some dollars.
- ☐ Do make insurance a part of your buying decision when shopping for a new car. Be aware, for example, that companies make surcharges for cars which are most frequently involved in accidents and losses and are very costly to repair.
- ☐ Don't forget the discounts offered on safer, less costly cars which are involved in fewer accidents and losses.

CHAPTER 14

HOW TO PROTECT YOURSELF AFTER THE ACCIDENT

All your precautions to protect the investment you have in your car and save yourself from ruin come to their big test the second you collide with another car. Instantly there is an implied threat against the sanctity of your possessions and your wealth itself. You are confused about what to do, and your heart is sinking with the fear that you now will be subjected to endless, wasted hours and days of long rituals with courts, lawyers and insurance companies. And for a passing second you wonder how long you will be without your car.

It's an emotional moment as well. A San Francisco driver could hardly keep his voice steady when he announced to his insurance agent that his brand new luxury car—which was equipped with expensive water bumpers to protect both its front and rear—had been hit broadside.

No matter how safe a driver you might be, accidents do come to us all. For example, there was one Nebraska woman in her seventies who had driven for years without so much as a dented fender. But, in the space of an hour, she smashed into 10 parked cars, came around the block and smashed for a second time into one of the same cars. The insurance paid for 11 separate accidents.

One young driver, proud of his newly acquired driving license after passing his tests with flying colors, picked up his auto insurance policy from his agent. On the way home he collided with the entrance gateway of a motel. Given a rental car for temporary use while his car was being repaired, he wrecked it in a parking lot. When he finally picked up his own repaired vehicle, he hit a light pole backing out of the repair shop. Possibly he was just trying too hard.

Once an accident does happen, be careful. There's opportunity aplenty to make mistakes at this point. You may be able to avoid a lot of trouble and extra expense if you have committed to memory beforehand the various steps you should follow. It's a good idea write them down in a notebook, and keep them in your glove compartment, together with some extra pages and a pencil.

Here are the first steps:

- Call the local police immediately (or have someone else make the call). If anyone is injured, give whatever help you can, but avoid moving the person so that you don't aggravate the injury. Covering an injured person with a blanket and making that person comfortable is usually as much as you can do.
- The police must be notified if the accident results in injury or damage that is more than minor. Most states require on-the-scene accident reports and many require written reports within a few days if the accident results in injury or death or property damage over a specified dollar amount. Furthermore, you may also need a report to support your own claim for damages.
- When talking to the police over the telephone, be sure to tell them as best you can how many people were injured and the kinds of injuries. The police will then notify the nearest medical unit or units. Do not try to contact a doctor or a hospital yourself because you may be wasting valuable time by trying to call one that is farther from the

accident than others. The police usually do a much better job by radio.
- Cooperate with the investigating officer and provide whatever information is required. But avoid making self-incriminating statements (not even so much as, "Sorry, it was my fault"). Stick to the facts.

There are a number of other steps you should take as well. If you're hurt, you may have to ask someone else to help. That's another reason why it's a good idea to have a written list of things to do.

- If another driver was involved, exchange information on driver's license numbers, license plate numbers, registration (owner of car) and insurance companies. Be sure to get the address and phone number of the other driver, as well as the names and addresses of all the passengers in the other car. Make notes about their injuries, if any, also if they say they are unhurt. Should a driver attempt to leave the scene before police arrive, write down the license plate number immediately.
- Record in your notebook the names and addresses of as many witnesses as possible, as well as the names and badge numbers of police officers or other emergency personnel.
- If you have a camera in your car, by all means take pictures showing the damage, the positions of the cars, skid marks and anything else that might document what happened.
- If you don't have a camera yourself, someone else may have one. Ask them to take pictures and arrange to get copies or the film.
- Make a sketch showing the positions of the cars before, during and after the collision. Your insurance company may want this kind of information even if yours was the only car involved.
- Take reasonable steps to protect your car from further damage. This might involve setting up flares, getting the car off the road, calling a tow truck. Have the car removed to a repair shop if necessary, but remember your insurance company will probably want to have an adjuster inspect it and appraise the damage before you order repair work done.

- Ask the investigating officer where you can obtain a copy of the police report later on. Sometimes you may have to order the report by mail, and perhaps pay a fee.
- If you run into an unattended vehicle or an object (even a farm animal), try to find the owner. Failing that, leave a note containing your own name, address and phone number.

Even if the other driver was at fault, think twice before accepting any offer of money to pay for your injury or car damage. The damages might be more extensive than you thought, or after-effects from an injury may show up only later. Your acceptance of cash may be a bar to any later claim against the other driver.

How to Put Your Insurance to Work

If your car is involved in an accident—or damaged in some other way such as by fire, flood or vandalism—or stolen, you should notify your insurance company as promptly as possible. Indeed, the terms of your policy probably require this.

Usually, your first step will be a call to your insurance representative. If you have a claim that's covered by your own policy, the insurance spokesperson can tell you how to proceed and what documents you will need to support your claim. If you have an agent, he may handle many of the details for you.

Even if you have a loss while you're away from home, say in Florida or California for a winter vacation, put in a call to your own insurance representative unless you have different instructions from your insurer. Your agent will want some details and a phone number where you can be reached. A local representative of your insurance company probably will then follow up on the matter.

If your car is damaged, you should file a claim with your insurance company if the estimated cost of repairs exceeds the amount of your deductible. Even if you file a claim with the insurer of the driver who was at fault, you should notify your own company. If you have collision coverage, its terms provide for your insurer to pay for the damages, minus the deductible, and then to seek repayment from the other driver or his or her insurer. Possibly, you'll get back the amount of the deductible— an amount you've had to pay out of pocket.

Your company may require a "proof of loss" form, as well as any documents relating to your claim, such as medical and auto repair bills and a copy of the police report.

If you have an accident caused by a motorist who is uninsured, or by a hit-and-run driver, file a claim with your own insurance company under the uninsured motorist provision of your policy.

You may have no-fault auto insurance or live in a state that requires it. If you do, claims for injuries sustained by you or members of your household in an accident should be filed with your own insurer, no matter who caused the accident. Claims for injuries suffered by an occupant of your car or by a pedestrian you hit will be handled by your insurer in some cases, by the injured person's company in others. Any questions about your proper course of action should be directed to your own insurance representative.

Once again, be sure to keep a careful accounting, and records, of any expenses that may be reimbursable under your policy. Remember, for example, that your no-fault insurer usually will pay not only your medical and hospital expenses, but such other costs as lost wages and at least part of your outlay if you have to hire a temporary housekeeper.

You'll be expected to cooperate with your insurance company in its investigation, settlement or defense of any claim, and to turn over to the company copies of any legal papers you receive in connection with your loss. Your insurer will represent you if a claim is brought against you and defend you if you are sued, but you'll be expected to provide the company with whatever information is needed and to testify in court if necessary. Be sure you always give your insurance company full and truthful information. Any attempt to mislead or falsify not only could jeopardize your defense, but it could invalidate your coverage and give your insurer a legal right to withdraw from the case.

Although perhaps not a model driver, a Minnesota policyholder of one large casualty company unquestionably was the model of honesty. When the claims representative asked him why he hit the other car, the man replied: "I didn't see the red light or the other car, because the pot smoke was too thick in my car."

Your auto insurance company is not responsible for representing you in connection with a claim that you, as the injured party, may file against another driver or that driver's insurer. If you have a problem with such a claim, you may want to consult a lawyer. However, even if you think you have a clear-cut case, don't neglect to report the accident to your own insurer. Even though you don't expect to, you may have to file a claim with your company to cover the cost of auto repairs or a portion of your medical bills. Or, the other driver might unexpectedly file a counterclaim against you.

Aid for the Innocent Victim

If you are injured by a hit-and-run or uninsured driver, there is help for you in some states even if you have no car—and therefore no auto insurance of your own. Several states maintain funds to aid residents without auto insurance.

In New York State, for example, no-fault personal injury benefits are available to a qualified resident who is injured within the state by an uninsured or hit-and-run driver or while riding in an uninsured car. The victim has 90 days to file an uninsured motorist claim with the state's Motor Vehicle Accident Indemnification Corporation, which is authorized to pay for up to $50,000 of medical bills, $50,000 in case of a fatality, and even up to $10,000 in "pain and suffering" benefits.

The driver, if he or she can be identified, doesn't get off scot-free. He'll not only face the loss of his driving privileges, but will be expected to make restitution, with interest.

Michigan, New Jersey and North Dakota have unsatisfied judgment funds to pay accident victims at least a portion of uncollectable judgments against negligent motorists. Those states, too, will suspend the driver's license and seek restitution.

If you live in a state with a no-fault auto insurance law and you're the uninsured victim of a hit-and-run or an uninsured driver, you may qualify for such benefits. Get in touch with your state insurance department, which usually is located in the capital city, or with your lawyer.

CONSUMER TIPS

For Your Glove Compartment

Knowing what to do in case of an accident. Knowing how to file a claim. These are the keys to getting the best value for your auto insurance dollars. Keep this checklist in your car at all times.

WHAT TO DO AFTER AN ACCIDENT:
- ☐ Get help for the injured.
- ☐ Call police.
- ☐ Provide accurate information to investigating officer.
- ☐ Exchange information with other driver(s) involved.
- ☐ Note data on...
    - ☐ Witnesses
    - ☐ Police officers
    - ☐ Other emergency personnel
- ☐ Take photos.
- ☐ Make a sketch of the accident situation.
- ☐ Protect your car from further damage.
- ☐ Ask police how and where you can obtain a copy of accident report.
- ☐ Leave a note if unable to locate owner of other car, animal or object.

HOW TO FILE A CLAIM:
- ☐ Notify your insurance agent or company representative as soon as possible. Ask him/her questions such as:
    - ☐ How to proceed?
    - ☐ What forms or documents are needed?
    - ☐ And other questions or concerns you have.
- ☐ Keep records of all your accident-related expenses.
- ☐ Supply information as company requires.
- ☐ Keep copies of all of your own paperwork for your own files.

PART THREE

Your Life And Health

CHAPTER 15

PROTECTING YOUR ESTATE

Sudden death is an ever-present risk that all of us face. You can live right and not take any chances. But you still can't say it won't happen to you.

You'll want to protect your dependents against the worst financial consequences if suddenly you shouldn't be there. There'll still be bills to pay, taxes, daily needs to be met, educations to pay for. If you die of a prolonged illness, there'll be a legacy of heavy hospital and doctor bills for the survivors to bear.

If you are a wife in these days when two paychecks are the norm, your loss could cripple the family's finances. Over and above normal expenses, your husband might have to hire extra help, for the children and the house.

There's only one way to guard financially against the ultimate risk of unexpected death. That's to have plenty of money. And if you don't have that, you'd better have a life insurance program that can provide at least some replacement for lost income.

Most men don't think of insurance on their wives as essential, but Gary Belson never expected that he would find himself a widower at the age of 33 with three small children, 7, 5 and 2 years old. Gary's wife Martha had cancer, and the time between the diagnosis of the disease and Martha's death was amazingly short. In the first moments of grief Gary wanted to reach out to one of the grandmothers for help in raising his children, but his mother wasn't in good enough health to manage the three youngsters, and Martha's mother was too overcome with grief to deal with the children.

Then Gary remembered that Martha's life insurance was substantial enough to pay for a housekeeper. Although the two of them had shared the responsibilities of home and children, Gary couldn't have handled everything alone without the help Martha's insurance provided.

The question of what would happen if a loved one died may be painful to answer, and good sense demands that you address it carefully. There are always emotions to deal with when someone dies, but life insurance can ease the many devastating practical problems that always follow.

Savings Features

Besides what are called death benefits, insurance policies help furnish cash for a major cost such as college tuition or to provide retirement income. The price of a college education has continued to rise, and inflation continues to cut into the value of pensions. For such needs, a five- or six-figure insurance policy could be an enormous help. For most people—you may be one of them—it would be almost impossible to save up such an amount.

Life insurance is a mystery to many people. But it doesn't have to be. As with other insurance, there are just a few basics you need to know about life insurance in order to spend your money wisely.

There is no such thing as a standard life insurance policy. The words and the appearance of each policy are determined by the company that issues it. Policies cover all the conditions under which claims can be paid as well as the amount of payment and the options your beneficiary will have when deciding how to receive payment.

Life policies today are much easier to read than they used to be. If you still have some questions, ask your insurance company representative or agent. Remember the company can be held only to what it puts in writing.

When you buy an insurance policy, you may be asked to take a medical exam, and you must fill out an application form. The insurance company will check the statements you make on the application to be sure that you've given your correct age, you live where you say you do, and all the other information you've given is accurate.

When you buy your policy, you will be billed for your premium monthly or annually or on some other agreed-upon basis. You'll have a grace period—usually 31 days—to pay the premium. If you don't pay in that time—for any reason—the policy will lapse. This means the insurance will no longer be in effect.

If you decide to stop paying your premiums, you can ask for the cash value of your policy, if any. You also can ask whether you can trade in your policy for paid-up coverage with a smaller face value or a policy having the same face value for a shorter period of time.

If you do not cash or trade in your policy, you'll usually be able to reinstate it within three to five years unless it's term insurance and the term has expired. But, to reinstate it, you'll have to pay whatever you owe in premiums plus interest for the time you haven't paid the bill. You'll also have to prove to the insurance company that you still qualify for insurance.

Here's a breakdown of the several types of life insurance policies available.

Term Insurance

With term insurance you receive a payment only if catastrophe strikes. Traditionally, the person who bought term insurance has been the father of a young family who wanted to guarantee that his wife and children would have money to cover their basic living expenses if he dies. Now, the buyer of term insurance often is a single mother who wants to provide for her children or a single person who supports aging relatives. Homeowners often have term insurance to guarantee that their mortgages will be paid up even if they die.

Term insurance is temporary; it protects the buyer for a term or a certain number of years—say 5 or 10 or 20 years or until the buyer is 65. If you buy a term insurance policy and you die before the end of the term, your beneficiary will receive the face value of the policy—that is, the number of dollars stated on the face of the policy. If you live past the end of that term, your policy has no value. It will not provide you with any kind of income, and you can't borrow against it. Term insurance has one big advantage for you. It costs less than any other life insurance you can buy. It gives you the largest amount of pure insurance protection for the smallest premium.

Variations of Term Insurance

There are several different types of term insurance.

Level term insurance has the same face value for the term of the policy. If you buy a $25,000 policy, it will have that face value for the duration of the term.

Decreasing term insurance, on the other hand, has a face value that keeps getting smaller every month or every year, while the premium stays the same. The persons who buy decreasing term insurance usually do so because they believe the amount of money they need to protect their beneficiaries will be smaller as the years go by. They frequently use this type insurance to cover the balance on their mortgage as it declines.

For example, say Bill and Sally Lewis have two small children. Bill has bought a decreasing term life insurance policy. If Bill dies when the children are still small, Sally will have a larger amount of money to pay for her needs and the children's. As the years go on, the amount of coverage will decrease. But, as the children get older, they will have fewer remaining years of school to finance. If Bill dies after the children have finished school, Sally will have less money, but by that time the children probably will be capable of supporting themselves.

Increasing term insurance, as the name implies, increases in value as the term goes on. A policy's face value may increase, say, 10 percent each year. Persons who buy this type of insurance usually do so because they want to make sure that inflation doesn't eat away at the amount they've provided for their family.

As your income grows and you can afford the higher premium for whole life or an endowment policy, you may want to change your term insurance policy into that type of insurance. If your term insurance is convertible, you can make the change without having to take a medical exam. If your term insurance is unconvertible, you won't be able to change it. So, at the time you buy your policy, be sure to ask whether it is convertible or not.

Term insurance can also be renewable or nonrenewable. If it's renewable, when the term of the policy ends you can renew it for another term of 5 or 10 or 20 years without taking a medical exam. You can keep renewing the policy up to a certain age. For some companies, that age is 55; for others, it's 60 or 65. If the original term policy is nonrenewable, you'll have to look for another policy when your original term expires.

Even if your policy is renewable, you'll have to pay a higher premium for your next term policy. The reason for the higher premium is that you're older and your chances of dying before the end of the second term are higher than they were before.

Insurance for the Whole of Your Life

Whole life insurance is permanent insurance, usually bought in units of $1,000. As a policyholder, you'll pay premiums for a certain number of years or for the rest of your life. The premiums will stay the same from year to year. How much you pay depends on your age—and the amount of coverage—when you take out the policy. The younger you are at that time, the lower the premiums will be. When you die, your beneficiary will receive the face value of the policy. A whole life policy also contains a savings feature through which it builds up a cash value for you.

Cash value is the money you would receive if you closed out your policy before you die. Most policies don't have any cash value the first or second year. If you borrow any part of the cash value, then called the loan value, you will have to pay interest on it, because you are borrowing your company's assets. These are supposed to be earning interest for all its policyholders. There's one big advantage to borrowing on your life insurance. The interest rate is usually well below what you'd pay to a bank or some other lender. There's another advantage, too. You can pay back the loan at one time or in several payments. You don't even have to pay back the loan. If you die before it's paid, your beneficiary will receive the amount that's left when the money you borrowed, plus any unpaid interest, is subtracted from the face value of the policy.

Endowment Policies

Endowment policies often are bought to pile up funds to pay college costs or to meet some other major financial goal. They're usually written for a certain period such as 10, 15, 20 or 30 years. At the end of that period, the face value of the policy is paid to the owner.

Let's say that when you're 25 years old, you buy a 20-year endowment policy with a face value of $20,000. When you're 45, you'll receive the $20,000 as a lump sum or, if you prefer, as monthly income. If you die before the 20 years are up, your beneficiary will receive the $20,000. Endowment policies are more expensive than whole life policies and lose some of their attractiveness in periods when the rate of inflation is high.

Special Types of Life Plans

Family income policies are a combination of whole life and decreasing term insurance, which runs for 10, 15 or 20 years, as the policyholder decides. If you die before the end of the protection period, the face value of the whole life part of the policy is available to the beneficiary immediately. The decreasing term part of the insurance provides a monthly income until the end of the protection period. If you live to the end of the protection period, the term insurance ends, the premium is reduced to cover the whole life part only, and the policyholder continues to hold the whole life policy.

A family plan policy is also a combination of whole life and term protection, but it covers all members of the family. The wage earner has the largest amount of insurance, which is whole life. The spouse and each of the children have a smaller amount, but their protection is term insurance. Children, when they are 15 days old, are automatically included in the policy. If you buy a family plan policy instead of individual policies you will save on service charges. Moreover, your premium payments will be more convenient.

If you buy a family plan policy, keep in mind that the primary purpose of life insurance is to provide funds to make up for loss of income. For that reason, the wage earner should carry by far the most insurance, and the homemaker should carry enough to pay for the services of a household manager until the children are old enough to fend for themselves. But children usually are covered by only enough insurance to pay their funeral expenses.

Mortgage cancellation insurance is a policy designed to pay off the remaining balance on your home mortgage, thus leaving your dependents with a deed, not a debt. It is usually a declining term policy, adjusted to the declining balance on your mortgage. It's one way of guaranteeing your family won't be put out on the street if you die.

Group life insurance is a benefit offered to employees of many companies or to members of unions or professional organizations. The premiums for group life insurance are lower than for an individual policy because group life term insurance is cheaper for the insurance company to administer. The employer or the secretary of the group keeps the records and pays the premiums with one check. All or part of the premium often is paid by the employer. So group insurance where you work is usually your best life insurance bargain. Recently, a few companies have begun offering group spouse coverage with which an employee can insure a spouse for the same amount that he or she has.

When a person leaves a job or quits an organization, he or she cannot continue in the group insurance plan. But, usually, the group coverage can be converted to a similar individual plan, with the premium based on the person's age at the time.

Variable life insurance invests your money in stocks, bonds and other securities. Many companies put a certain amount, say 25 percent, into guaranteed securities like government bonds. The other 75 percent goes into a mutual fund or stocks that you specify. The mutual fund and stocks rise and fall in value with the stock market. If the securities grow in value, the death benefit and the cash value of your variable life insurance policy rise, too. If the value of the securities falls, the death benefit and the cash value decrease, but only to a limit that you agree to when you buy the policy. You can also get a variable annuity, where the monthly payout to you is pegged to the market value of securities in which your money is invested.

Universal life is a new kind of insurance policy first issued in 1980. While a universal life policy offers the permanent lifetime protection of a traditional whole life policy, it also pays a higher return on the savings portion of your investment than do whole life policies. The features of universal life policies differ from company to company, but with a number of policies you can raise or lower the face value and your premiums, according to your needs. You may have a choice of investment options. You also can choose from several payment options. The objective of universal life is to take advantage of current high returns on investment for your benefit, thus making these life policies competitive with other savings vehicles.

A feature of some universal life policies is a loan option, which has been offered as a tax-free return on the policyholder's investment. But, the tax status of such policies has not been finally resolved. So, if you are interested in a particular universal life policy, be sure to ask for the latest tax information on it.

Special Coverages

There are other special coverages called riders that you can add to a basic insurance policy. For instance:

Double Indemnity—You can add an accidental death benefit to your life insurance policy, so that your beneficiary will receive twice the face amount of the life policy if you die because of an accident. This benefit is called double indemnity. For a further premium, the basic coverage may be tripled. This benefit is called triple indemnity.

Most companies make the accidental death benefit available to people between ages 10 and 65. Accidental death must occur within a certain time after the injury, usually 90 days. Accidental death benefit rules out certain causes of death such as war and riots. And, of course, the policy does not pay if it is determined that a supposedly accidental death was not really accidental. The murder movie "Double Indemnity" was based on this type of situation.

Don't let double indemnity mislead you. Some people believe that they will die in an accident, and they pay the premium for double indemnity and then tell themselves they have twice the amount of insurance they really have. But a typical case shows the fallacy of this line of reasoning and the value of carrying enough basic insurance to meet your family's needs, without depending on special provisions such as double indemnity.

Harry Shaw's father and grandfather died in transportation accidents before they were 45. Superstitiously, Harry believed he would share the same fate. When Harry turned 40, he added a triple indemnity rider to his policy. His two children were 10 and 12 at the time, and $90,000 sounded like a lot of money to Harry—certainly more than enough to pay for the children's education and to keep the household going until they went off to college.

Harry did die two years later but the cause of death was a massive coronary, and the $30,000 his wife received wasn't nearly enough to maintain the family and see the children through college.

Waiver-of-Premium—Waiver-of-premium is a low-cost way of making certain that your insurance premium will be paid if you become totally disabled. Total disability is disability that prevents you from doing any work for which you are paid.

Companies differ on how long you must be disabled before your premiums will be paid for you. Some firms waive premiums after six months, others after four, and a growing number are agreeing to waive premiums after 90 days of total disability.

Companies will issue waiver of premium to people who are less than 60 years old. To have premiums waived, you must become disabled before 60.

Waiver-of-premium is an option that provides payment of premiums until the end of the term. Some waivers allow persons who become disabled to turn term insurance into a whole life policy.

Remember that waiver-of-premium covers only premiums; it does not provide disabled persons with any kind of income. However, it can be worth the additional charge you pay for it.

Disability Income Rider—With some life insurance policies, you can purchase a disability income rider to protect you against financial crisis resulting from disability. The disability must be total, and you must become disabled before you are 60 years old. For every $10,000 of life insurance you hold, you can receive a stipulated amount of monthly income, say $50 or $100, depending on the size of the premium you pay. Most companies limit the amount you can receive each month to $500 and require that you be totally disabled for six months before you begin receiving benefits.

Some life policies also offer a cost-of-living rider which allows an increase in the face amount of the policy, based on a measure such as the government's Consumer Price Index, without requiring evidence of insurability.

Guaranteed Insurability Option—If you take the guaranteed insurability option, you'll be able to buy more insurance in the future without having another medical exam. This option is offered to persons under 40 who buy a whole life or endowment policy. It allows them to buy more of the same insurance every three years without demonstrating that they are physically qualified.

If you're worried that a health problem might prevent you from buying more insurance in the future, this option may give you an added sense of security.

Annuities

Annuities are plans that provide you income for the rest of your life. It's possible to turn your life insurance policy into an annuity when you retire, but you'll have to read your policy to see what provisions govern this change. The types of annuities described below differ in what happens to what's left of your money after you die.

Payments from a straight life annuity end with your death. If you have no dependent or if you've provided for your dependents through other financial plans, you may be interested in a straight life annuity.

But if you'd like a plan that will furnish payments for your dependents as well as for yourself, you might want a life annuity with a guaranteed period of payment, usually 10 or 20 years. If you die before the end of the guaranteed period, your beneficiary will receive income payments until the end of that period. Some companies allow beneficiaries to withdraw a lump sum from the plan as a substitute for monthly payments.

If you have an installment refund annuity and die before you've received as much money as you've paid in, your beneficiary will receive income until the total payments equal all the money that you've paid.

A cash refund annuity will allow your beneficiary to receive the balance in one lump sum.

Just remember, once you've begun receiving annuity payments, you won't be able to get the balance of what's owed you in one check.

If you decide to retire at 55 or 60 instead of at 65, talk to your company representative or agent. You can probably start your annuity payments earlier than planned, but your payments will be considerably smaller than they would be if you retired at 65, because you will have paid in less money and you will be receiving income for a longer period.

Policy Dividends

You will receive policy dividends on the premiums you pay if you have what is called a participating policy. This way you share the costs of providing the insurance. If the premiums the company received and the interest from its assets are more than what was needed to run the company and cover its claims, the company will give you back some of your money in the form of dividends. Most participating policies are sold by mutual life insurance companies, which are companies that are owned by the policyholders.

Stock life insurance companies are companies owned by the people who bought shares of stock in a company either when it was first incorporated or later on. Although many pay policy dividends, stock companies generally sell nonparticipating policies, which do not pay dividends.

Initial premiums for participating policies tend to be higher than those for policies that do not pay a dividend, but they may be reduced—sometimes considerably—by policy dividends.

How Can You Use Your Policy Dividends?

If you decide on a participating policy, you can take your dividends in a number of ways. You can receive them in cash, you can use them to pay part of your premiums, you can buy more insurance with them, or you can let them gather interest. Be sure you find out what rate of interest you are receiving on your dividends. At certain times, you'll find that you'll get a greater investment return elsewhere.

Are Dividends Taxed?

You don't pay income tax on dividends, but if you let them accumulate interest, you will be taxed on that interest just as you are taxed on the money you keep in a savings account in a bank.

There are other times when policyholders or beneficiaries are taxed on money they receive from insurance policies. Death payments are not taxed unless they are part of a six-figure estate, with the six figures varying from Maine to California. In such a case, death payments over a certain amount are subject to tax, along with all the other items in the estate. Also, if payments from a life insurance policy are received in installments rather than all at once, each installment contains some interest, and that interest is taxed.

Shopping for Life Insurance

An adage says, "Life insurance isn't bought; it's sold." The adage probably arose from the old days when some salesmen would just say "Sign here," and a buyer would do just that, without stopping to ask questions. The fact is you should shop for life insurance the way you shop for anything else that costs a considerable amount of money and is important to your family's welfare.

First of all, understand your particular needs and know what you want your insurance policy to do. Are you trying to provide an income for your dependents in case you die when they are unable to support themselves? Do you want to build up a fund that you can reach for when you need a large sum of money, say, for your children's college expenses? Or do you want to put some money aside for your retirement?

Looking for a Policy

Once you have established what your needs are, you should have an idea of the kind of policy you want. See a company representative or an insurance agent and tell him or her what you're looking for.

Let's say you think you want a nonparticipating term life insurance policy for 15 years. Find out what your insurance representative can offer you. Ask what the premium for this policy is. Compare it with others. See another company representative or agent and run through the questions again. Perhaps you may want to talk to a third company representative or agent, too. This will take time, but if you were buying a new car, wouldn't you expect to spend several weekends deciding what kind of car you wanted and where you could get the most for your money?

As with a car, price shouldn't be the only basis on which you make a decision. You are unlikely to find exactly the same policy offered by two companies. Each company, in trying to attract customers, puts together a policy with different features. If you were buying a car and operating on a tight budget, you wouldn't be taken in by a model with lots of gadgets. So, don't be talked into some features in an insurance policy that may be more suitable for someone with a higher income. But if you find a company representative or agent who tries to understand your situation and who goes out of his or her way to explain the options and to put together a life insurance program tailored to your needs, listen carefully.

Then, too, don't just buy a policy and forget about it. Despite the importance of insurance policies, people often are very vague about what they contain or how much they are worth.

Lila James lost her husband Dave a little over a year ago. She thought they had allowed the insurance policy—which they had bought in 1958—to lapse, so she didn't contact the insurance company. Luckily, she ran into Steve Carter, the agent from whom her husband had bought the policy, and Steve asked if Lila was receiving her insurance payments regularly. When she replied that the policy had lapsed, Steve decided to check and found that the policy was indeed in effect. Lila soon began receiving income from the insurance company.

As your needs change, call on your life insurance agent or company representative and ask for help to determine if your policy is still the best one for you to have.

Another item to keep in mind is your designated beneficiary, which you may want to change at some time. You certainly wouldn't want to be like the soldier who named his fiancee as his beneficiary on his GI life insurance. When he was killed in Vietnam, a "Dear John" letter was found among his effects. It informed him that his beloved had married another man. Because he hadn't changed his beneficiary, the other man's bride received all the proceeds from his life insurance. An ironic—and unintended—wedding present.

Products to Match Your Interests and Needs

In trying to keep up with the times, life insurance companies are now offering products that match buyers' changing interests and needs. For example, as an encouragement to Americans to pay more attention to their health, some companies offer discounts to people who don't smoke or who take regular medical exams and participate in such physical fitness programs as aerobic exercises. If you participate in such a program, ask your agent if the company you are dealing with offers such a discount. If it doesn't or if you can't get satisfactory answers to your questions, call another agent.

If you're a senior citizen with a life insurance policy you bought years ago when your children were young, you certainly ought to take another look at the policy as well as your own needs. For example, if you have a beneficiary who needs the money when you pass away, you probably should leave the policy just as it is. However if your beneficiary is financially comfortable while you're finding that your Social Security benefits don't cover as many of your expenses as you'd like, you should see about trading in your policy for a lump sum. The money will help pay your expenses for the rest of your years.

Instead of a lump sum, you might prefer to take the money coming to you in monthly retirement checks. But there's a drawback to this. If you're ill, it might not pay to change your policy to retirement income because what you receive in monthly checks may end up being quite a bit less than what your beneficiary would receive if you die. The monthly income option will pay as much or more than the face value of your policy if you outlive the statistical life expectancy your monthly benefit is based on.

Finally, if you're still paying premiums on your policy and are finding it hard to meet those payments, you ought to talk to your company representative or insurance agent. You can stop paying the premiums and have a paid-up policy with a smaller face value than that on the policy you bought. Your agent will give you all the information you need in order to decide what would be the best step for you to take.

Years ago, people considered life insurance a taboo topic because they were afraid to talk about any matters related to death. Now, thanks to greater sophistication in several fields, they're seeing life insurance for what it is: a way to prevent a financial disaster if a major contributor to the family's economic situation dies, and a sound way to build or create a financial estate.

CONSUMER TIPS

Look Out For Yourself

Think of your life as an asset. Consider the condition of your financial security in these kinds of situations:

- ☐ Half your regular family income is no longer available because a breadwinner has died or become disabled.
- ☐ Your twins are going to be ready for college just two years from now.

Review your life insurance policy to see what it contains.

Call your agent to ask questions about anything you don't understand.

Be certain to recognize that a life insurance policy is a legal contract and that the only conditions under which claims can be paid are those which are written into the contract.

Check periodically to see whether the beneficiary named in your policy is still your choice.

Decide whether your present policy is the type which best suits your current needs. Should you:

- ☐ Ask the company to give you the cash value?
- ☐ Trade your policy in for a paid-up policy with a smaller face value or a policy of the same face value for a shorter period of time?

Find out just what type of life insurance your policy represents. Do you have or should you consider...

TERM INSURANCE?

| ☐ Level term? | ☐ Increasing term? | ☐ Decreasing term? |

Is your term insurance:

| ☐ Convertible? | ☐ Non-convertible? |
| ☐ Renewable? | ☐ Non-renewable? |

WHOLE LIFE INSURANCE?

Do you know its: ☐ Cash value?   ☐ Loan value?

Protecting Your Estate

ENDOWMENT LIFE?
OTHER TYPES?
- ☐ Family income policy?
- ☐ Family plan policy?
- ☐ Mortgage cancellation insurance?
- ☐ Group life insurance?
- ☐ Variable life insurance?
- ☐ Universal life insurance?

Consider the special features of your policy or whether you wish to add such features as:

ACCIDENTAL DEATH BENEFIT
- ☐ Double indemnity
- ☐ Triple indemnity
- ☐ Waiver of premium
- ☐ Disability income rider
- ☐ Guaranteed insurability option

ANNUITIES
- ☐ Straight life annuity
- ☐ Life annuity with a guaranteed period of payment
- ☐ Installment refund annuity
- ☐ Cash refund annuity

FIND OUT WHETHER YOU HAVE A:
- ☐ Participating policy
- ☐ Non-participating policy

If you do not yet have a life insurance policy or are reassessing your needs, keep these tips in mind:
- ☐ Know what you want your insurance policy to do for you.
- ☐ Have an idea of the kind of policy you want.
- ☐ Contact several agents or company representatives and tell them what you're looking for.
- ☐ Find out what each representative can offer you. Ask what the premium will be.

COMPARE THE POLICIES:
- ☐ Coverages—amount and type
- ☐ Features
- ☐ Options

    Ask questions and listen carefully to the answers

    Inquire about any discounts for which you may be eligible:
    - ☐ Non-smokers discount
    - ☐ Discount for people who have regular medical exams and take part in a physical fitness program

CHAPTER 16

WHEN SICKNESS STRIKES AT YOUR SAVINGS

Not so long ago a Florida man was filing a claim for his wife's pregnancy, and when he came to a box on the form asking whether the claim was related to an accident in the home he wrote: "Yes—in a fit of emotion!" Whether it's a real accident like falling down the stairs, or something routine like having a baby, when your body needs medical attention the costs to your savings can be unbelievably draining. Obviously, the answer is to protect yourself ahead of time with the right kinds of medical insurance.

A lot of people ask how much hospitalization and surgical coverage they should have. To judge that, you need to keep in mind that health care costs have risen far more rapidly in recent years than other costs of living. The Consumer Price Index, published by the U.S. Department of Labor, shows that the overall cost of medical care rose 230 percent in the 15 years between 1967 and the middle of 1982. The cost of physicians' services went up 229 percent, and the per-day rate for a hospital room leaped by 461 percent. In contrast, food and beverages rose 180 percent, and clothing prices rose 91 percent. The index of all consumer items went up 191 percent in the same period.

Thus it's essential that you have health insurance that will protect adequately your savings and financial health without duplicating your other coverages.

Most people know of cases like that of the Pritchetts. They had saved up a nice retirement nest egg and were just about ready to retire when John Pritchett was stricken with lung cancer. After a series of operations to remove portions of his lung, and a long hospital stay, John was released. But because he was not yet 65, he wasn't eligible for Medicare, and his limited form of health insurance only covered about a third of his expenses. His retirement money was wiped out, and he had little chance of building it back because his income for the next few years would have to be used to pay off his medical bills. His wife Mary even had to find work at the age of 60 to help. With the right form of health insurance, the financial problem would have been taken care of and the Pritchetts could have had the retirement they had dreamed of and saved for.

To avoid the Pritchetts' fate you need to plan early. There are various types of health and medical coverage you should learn about: group plans, hospitalization and major medical policies, health maintenance organizations, health care contractors and Medicare/Medicaid. From their differing features, you can judge what's best for you.

Group Health Insurance

Group plans, which may be offered through your employer, alumni or professional association, or fraternal group, are usually very good. They are cheaper for you than other plans not only because the employer may pay part or all of the premium, but because the insurance company charges less since furnishing insurance for a group requires less paper work than providing a lot of individual policies. Then, too, the policies usually are purchased by a personnel or financial officer who spends more time studying insurance than most people do, and therefore can make a better-informed choice.

Before you do anything else, you should see if you can qualify for a group plan. If you contemplate leaving your job for another one, you should investigate the kind of insurance you'll be able to obtain through your new employer. If that plan is not as good as the one you currently have, ask if you can convert your existing group plan into individual coverage. But your premium for an individual policy normally will be higher than the premium for the group plan. If you're fired or laid off, see how long your coverage will continue or if it can be converted to an individual policy. If necessary, find a temporary policy to take care of you until you get another job and are under group coverage again. Before you turn 65, try to convert your group health insurance into an individual plan that will supplement the Medicare coverage you will receive from the government.

If you can't buy group coverage, talk to the representatives of companies in your area who sell health insurance. Compare coverages and costs.

Health Maintenance Organizations

As an alternative to regular health insurance, check the yellow pages or the local medical association to see if there is a health maintenance organization (HMO) or health care service contractor in your area. Insurance companies pay you after you've received treatment, but HMOs and health care service contractors provide you treatment on a prepaid basis as you need it. Some even provide preventive health care.

Jean and Bob Abbott had been married for several years when they adopted two children, who were 3 and 5 years old. Jean and Bob had been very healthy and did not even have a family doctor. But, with the arrival of the children, the Abbotts were forced to pay more attention to health care. Fortunately, Bob was a faculty member at a large university, which sponsored an HMO. For a comparatively small amount of money, the Abbotts enrolled in the health group. Not only did they obtain the protection they needed for the times when one of them was ill, but Jean and Bob also got involved in a program of health care that was very instructive for a couple as new to parenthood as they were.

For a monthly fee, HMO members receive services ranging from regular checkups to major surgery. The surgery is performed by a doctor on the staff of the HMO clinic, the HMO doctor working out of his or her own office, or a surgeon on the staff of an appointed hospital. Most HMOs conduct numerous classes on managing stress, reducing weight, giving up smoking, planning a family, taking care of a baby and other aspects of health maintenance. They also encourage immediate treatment for minor ailments, which many people are prone to skip under traditional medical care.

Figures show that HMO members spend about half as much time in hospitals as persons with other kinds of health coverage. Although about 11 million people belong to HMOs, many people hesitate to join because they don't want to leave their own doctors. That has to be your personal decision.

Health Service Contractors

Health care service contractors either contract with hospitals and doctors to treat subscribers or they own hospitals and employ doctors and other staff. The Kaiser-Permanente medical facility in California is an example of a health care service owned by the sponsors. If you're a subscriber to a typical service and a participating physician treats you, your bill is paid in full. If an emergency occurs and you have to be treated by a nonparticipating doctor or at a nonparticipating hospital, the plan may reimburse you—but only for the amount that would be charged by its own doctors and hospital.

Each health care service contractor or HMO has its own requirements for membership, but undoubtedly you will have to live in the area where the plan is situated.

While companies that sell medical insurance also sell disability income insurance, health care service contractors and HMOs do not. So, if you purchase your health care from them, you may have to look elsewhere for a disability income policy.

In selecting an HMO or a health care service contractor just as much as in choosing a health insurance company, you will want to make sure the provider is financially sound and ready and able to provide services as you need them.

Basic Coverage

Basic coverage insures you against the costs of ordinary hospital care, surgery and doctor's services. Such a policy will pay for daily room and board in the hospital, routine nursing care and minor medical supplies. The policy also will pay for other hospital services such as laboratory tests and X-rays, anesthesia, use of operating room, drugs and medications and local ambulance service. The total you can receive for these additional services is often expressed as so many times the benefit for room and board. Most policies begin to pay benefits only if you've been in a hospital a certain number of days. If you're discharged before that number of days, you'll have to pay the costs of your hospital stay yourself.

Your policy usually lists the amounts of money you will receive if you have certain operations. For example, removal of a cyst may appear on the schedule as an $85 operation. Even if your surgeon charges $300, your policy will pay you only $85.

There are times when you will need a doctor's services, but when your illness will not be serious enough to warrant your going to a hospital. Doctor expense benefits also are provided by health insurance. Some policies may allow a certain amount for each visit. Other policies pay for the visits only after you have seen a doctor two or three times for the same illness. You will need to evaluate the adequacy of the offered coverage in relation to the realistic current costs of doctor visits and other medical care. You may decide that a basic hospital policy does not give you adequate benefits.

Major Medical Plans

A major medical policy provides coverage over and above that offered by most basic hospital-surgical and doctor bill plans. With a typical major medical policy, you pay a deductible of a certain amount, say 20 percent of the expenses, and anything else your policy does not cover, such as the television in your room.

Policies may put limits on the amounts of money to be paid for each illness or accident or for all the illnesses and accidents you will have in your lifetime. Deductibles may apply to each illness or accident or to the medical expenses of one year. Deductibles range from $25 to more than $10,000, and limits can be as low as $10,000 or as high as $1 million. Some plans may be unlimited.

Most major medical plans cover hospital and surgical expenses, as well as drugs, medicines, physicians' charges and nursing care. Many major medical plans require you to pay 20 percent of the cost up to a certain amount, $1,000 for instance. Beyond that, to the limit of the policy the company pays 100 percent. Other plans may pay 100 percent of eligible expenses after the policyholder has paid a specified amount, such as $1,500.

If you compare two policies that are identical except for the amount of the deductible, you'll usually find the policy with the higher deductible has the lower premium. So get as high a deductible as you can afford. Let's say you are purchasing a policy that has a deductible for each illness or accident. Think about how much you can really afford to pay for each episode of illness. If you determine that $200 won't tax your emergency fund too much, buy a policy with a deductible of at least that amount.

Bob Dunlap was self-employed, and because his income varied widely from year to year, he pared costs as much as he could. He decided to save money by not purchasing major medical insurance. Before serious illness struck Bob or his family, his next-door neighbor, Jeff Koenig, had a double coronary bypass operation. The Koenigs had major medical insurance with a $10,000 limit, but because the bypass operation cost $23,000, they ended up paying $13,000 out of their pocket. Bob Dunlap then realized it would be far wiser to buy a major medical policy than to risk incurring a bill like Koenig's, or larger.

If the policy has a maximum benefit for each sickness, it should be at least $25,000 per person for hospitalization in each benefit period and $10,000 for other medical bills per year to be worthwhile. If the policy has a lifetime maximum, make sure it's at least $250,000. These are about the lowest realistic limits for coverage that could be considered adequate under today's conditions. If you can afford it you should consider getting even higher coverage.

Disability Income Insurance

Disability income insurance pays part of your income when you can't work because of accident or illness. Everyone who depends on income from a job should have disability insurance in some form. However, many companies won't sell disability insurance to anyone who makes less than $18,000 a year, though a few companies have begun selling disability coverage to homemakers. You may have to have a medical exam if you are applying for a disability policy. If you are covered by Social Security, you are eligible for Social Security disability coverage. Some states as well have payroll deduction disability insurance plans. You can easily check the amount of protection these plans provide, either through your employer, the Social Security Administration, or your state insurance department.

For additional protection, you can buy either short-term or long-term disability insurance. Short-term insurance most commonly pays benefits for 13 to 52 weeks. Long-term insurance provides benefits for longer periods—two years or more, to age 65, or even for life. There is usually a waiting period of at least 30 days.

Under this type of coverage, you'll get a monthly check for a certain percentage of the salary you're earning at the time you become disabled. If you decide, for example, to buy a policy that will pay 60 percent of your salary and you become unable to work while you are earning $2,000 a month, you will receive $1,200 a month in disability payments.

Some policies make payments only to persons who are totally disabled. Some pay only for a limited period. To save yourself from misunderstanding and possible disappointment, be sure your policy makes clear the circumstances in which you will receive disability income, and that you know what these conditions are.

Disability insurance really can be a light of salvation at a dark time. Steve Buchanan was a healthy, hard-working father of two. For years he paid for disability insurance, often resenting the payments because he was in such good health. In fact, his premiums seemed ludicrous because his two daughters always seemed to be ill, mostly with minor ailments. And, Sally, his wife, had three major operations while the girls were still in elementary school and never was well enough to work full-time. When the girls were in college, Steve was totally disabled by a non-malignant brain tumor, something no one could have predicted. The family's lifestyle changed considerably, and the youngest daughter transferred from a college several hundred miles from home to an inexpensive community college nearby.

But the disability insurance Buchanan had wisely bought meant the family could stay in their home and Sally did not have to look for a full-time job.

Dental Insurance

A relatively new group coverage being adopted by some employers is dental insurance, which often pays 80 percent or more of charges for such ordinary dental procedures as extractions, fillings, bridges, dentures and the like. Other plans have scheduled fees for specific procedures. Because dental expenses are likely to be household budgetbusters—that is, somewhat unpredictable both as to timing and costs—this is a good coverage to have. It may be offered by the same insurer that underwrites your regular group health insurance. If your employer doesn't have it, you may find you can purchase an individual policy. But before you do, make certain you're not covered already in your company medical plan.

Specific Disease Policies

Here's one coverage you probably shouldn't have to buy. Insurance policies frequently are offered covering specific types of accidents or illnesses. Under certain circumstances, you might find such a policy desirable, but if you already have health insurance with a wide range of coverage, the extra policy shouldn't be necessary. If you are a concert pianist or a professional dancer, you might wish to have a special "dismemberment" policy on your hands or your legs, as some performers do.

What to Look For in a Policy

When you examine an individual or family policy, whether it's one you own or one you're thinking of buying, see if it specifically excludes certain pre-existing conditions, which are illnesses or injuries that began before the policy went into effect. Many policies have a one-to-two-year waiting period before you can be reimbursed for medical treatment of these illnesses.

Suppose you have a heart condition before you buy a policy. A few months later you have to go into the hospital because your condition worsens. The insurance company will not pay the cost of your hospital stay, since the condition existed before you bought the policy.

Even if you're dissatisfied with your current policy, you may be better off keeping it than buying another one that will make you wait a year or two until you are fully covered for all pre-existing conditions. If you're not sure about what you should do, talk to representatives of several health insurance companies and weigh and compare the advice they give you.

All health insurance policies have some exclusions, such as illnesses and injuries resulting from war or military service and on-the-job accidents covered by workers' compensation. Be sure you know what exclusions your policy contains.

You also should know your policy's terms of renewal. A guaranteed renewable policy can't be canceled until you reach 65, as long as you pay your premiums.

Payments begin a specified number of days after the onset or diagnosis of an injury or disease. The longer this waiting period, the lower the premium should be. The waiting period for a disability income policy could be as long as a year or two.

The time for which a disability policy will pay benefits is called the benefit period. It could be five years or until you reach 65. The longer the benefit period of the policy, the higher you can expect the premium to be.

When you buy a disability insurance policy, you can also purchase for an additional premium a cost-of-living rider, which will assure that your benefits will rise at a rate calculated to meet inflationary increases.

Your policy contains an incontestable clause, which means the insurance company that issued it won't be able to challenge the validity of the policy after it has been in effect for a certain number of years. This is especially important if you become disabled because of a condition for which you were treated before you bought the policy.

Medicare and Medicaid

As you grow older, your need for life insurance may decrease, but inevitably you will find that your need for health insurance is greater. In 1965 the federal government established the twin programs of Medicare and Medicaid. Medicare covers part of hospitalization, surgery and convalescence costs, mainly for people 65 or older. Medicaid is a health insurance plan for people out of work or whose income is low. Persons who are nearing the age of 65 and those who can't afford the cost of health and accident protection should contact their local Social Security office for Medicare or welfare office for Medicaid, respectively, to see if they qualify, and to apply for these programs.

When Ben Burroughs broke out in a rash shortly after his 65th birthday, he didn't think his condition was serious. But the rash made him uncomfortable, and he went to his family doctor. The doctor ordered some blood tests and before Burroughs could grasp the situation, he found himself in a hospital receiving chemotherapy for an acute blood condition. A week later the blood disease seemed to be controlled, but the next day Burroughs dropped into a coma and remained in that state for 54 days before he died. The Burroughs family was shocked by the swiftness with which a husband and father had gone from apparent good health to a critical condition. The bills for Burroughs' last illness were equally staggering; they came to $37,000. The family was still trying to figure out how they would manage to pay it, when they learned that the combination of Medicare and supplemental policies that Burroughs had bought paid for nearly every cent of his bill.

Medicare offers two kinds of coverage—Part A, hospitalization, and Part B, medical services. With Part A, for which no premium is charged, there is a deductible of $304. If you're sick more than once in a year Medicare pays full costs of hospital care for 60 days. These benefits include a semi-private room, all meals, regular nursing services, laboratory and X-ray fees, intensive care costs, operating and recovery room, anesthesia, drugs, casts, dressings, splints and in-hospital therapy. If you have to stay in the hospital longer than 60 days, you will have to pay $76 a day through the 90th day.

Medicare A also gives you a reserve of 60 days during your lifetime; you can use these days if any hospital stay lasts longer than 90 days. However, you will have to pay $152 for each reserve day you use, and once you use up the reserve it's gone. There's no way to replenish it.

After you spend three days or more in a hospital, you also will be eligible for 100 days in a nursing home. Medicare will pay full costs for the first 20 days, but you will have to pay $38 for each of the next 80 days.

Medicare Part A also provides for an unlimited number of home nursing visits by skilled personnel. These benefits can include therapy, skilled medical services and supplies and equipment. Medicare A doesn't pay for custodial nursing care. Custodial care involves only a patient's personal needs and can be provided by people without any professional skill or training. It includes bathing, dressing, feeding and helping people to get in or out of bed.

For Medicare Part B, you will have to pay a small monthly charge. After you pay a $75 deductible, Medicare Part B will pay for 80 percent of reasonable medical (other than hospital) charges; you will pay the remaining 20 percent. If your doctor charges more than the amount that Medicare has established as a reasonable cost, you will have to pay that additional charge.

Medicare Part B also pays toward diagnostic tests, artificial limbs, independent laboratory tests, certain ambulance services, radiology and pathology services, physical and speech therapy, oral surgery, limited out-patient psychiatric or chiropractic care, emergency room and out-patient clinical benefits.

Medicare B includes as well an unlimited number of home visits by skilled paramedics. Neither Medicare A nor B covers items such as private-duty nursing, routine checkups, eyeglasses, hearing aids, dental work, cosmetic surgery and prescription drugs not obtained in a hospital. The benefits for Medicare Part A and Part B do change from time to time, and you should check with your local Social Security office to be sure you have the most up-to-date information.

Since Medicare has deductibles and limits that leave you responsible for some of your medical costs (sometimes called the "Medigap"), you should buy a "wraparound" or Medicare supplement policy that will pay the deductibles and the parts of the charges you're responsible for. Some wraparound policies pay for only those items allowed by Medicare. Other policies pay also for out-of-hospital prescription drugs, medical appliances and equipment. They may also pay for hospital days beyond the number covered by Medicare and the co-payments for a long stay in a nursing home. You can buy a wraparound policy from your insurance company agent or representative. It is also available through some retirement associations.

After both their daughters married, Bert and Edith Ford sold their home in a Boston suburb and moved into a comfortable retirement community on Florida's Gulf Coast. A few months later, Edith suffered a massive stroke and required long-term hospitalization and nursing home care. Luckily, the Medicare supplement the Fords had purchased when they had turned 65 paid for the deductible and most of the costs of Edith's hospital and nursing home stay that Medicare did not cover. The money the Fords put into supplementary health insurance turned out to be one of the best investments they ever made.

If you are still earning money in your retirement you might consider, as an addition to a wraparound coverage, a hospital income policy that will provide you a certain amount of cash for every day you are hospitalized. You can use the money to pay the costs Medicare and the wraparound policy do not cover. Normally, you would receive anywhere from $10 to $80 or more a day, depending on the policy you buy and the premium you pay. If you buy a hospital income policy, you'll have to keep checking rising health costs, or you may find the fixed daily payment you elected to receive will no longer cover all the expenses. Hospital income policies are not as well geared to inflation as wraparound policies, since the latter pay a percentage of actual costs. Thus wraparound coverage goes up as the costs rise.

Why Does Health Care Cost So Much?

One of the worst-kept secrets in America these days is the unbounded leap in hospital costs. Anyone who has spent even a few days in a hospital asks why. Since higher costs mean higher premiums for medical insurance, policyholders are asking the same question. And so are insurance companies. For one thing, the medical profession has made some remarkable advances that a couple of decades ago seemed impossible. Now such things as transplants, dialysis machines, brain scans and fetal monitors are improving and prolonging lives. Yet, while their benefits are welcomed, their costs are often astronomical.

Considering the publicity attending these advances, it is understandable that people now expect better medical care. The response to consumer demands for more sophisticated treatment and services has been more tests, more equipment, more doctor visits, and more of nearly everything connected with hospital care. In fact, one of the growing criticisms of the current health system is that it concentrates too much on treating sickness and not enough on keeping people well.

What You Can Do to Lower Costs

Keeping well is one way you can hold down your medical costs. Health insurance policies carry a substantial premium, so there's little chance people will buy too much insurance. But when you are filing a group insurance claim, you will have to list other group policies that cover you. If all your costs are covered by one policy and you have duplicate coverages, you usually will not be reimbursed by both. Sometimes if one doesn't cover it all the other will take over. But, the maximum you will receive for any illness is no more than 100 percent of the total cost. Therefore, you should look at all your coverages carefully and save money by avoiding duplications of coverage. You can also save by emphasizing coverages for catastrophic illness, while paying your own bills as much as possible on small, affordable medical expenses.

Increasingly, doctors are telling us we have considerable control over our well-being. To minimize illness in your family, start taking better care of yourself and persuade your relatives to do a better job of looking after their health.

First of all, if you smoke, medical authorities advise that you try to stop. Newspapers and magazines are full of articles on the dangers of smoking. Thirty-one million people have stopped smoking, so it is possible for you, too.

Next, if you use alcohol, health experts recommend you control the amount you drink.

Maintain an optimum weight and watch what you eat—cut down the excess fats and sugars in your diet but be sure you are getting all the nutrition your body does need. One way to hold down weight is to exercise. Regular exercise is important for fitness and longevity. So set up an exercise program that's fun and stick with it. In a few weeks when you start noticing an attractive difference in the way you look, you'll be motivated to continue with the program.

Bob Jacobson, a 50-year-old six-footer who was overweight at 230 pounds, went for a medical checkup because he had been feeling sluggish. The doctor warned him that unless he took off 40 pounds and improved his fitness, he had a higher-than-average probability of heart attack. Bob and the doctor agreed on a program of diet and exercise, and he began going to a health club for a supervised workout on his lunch hour, stopping at a fruit stand for an apple or a banana instead of going to his favorite restaurant each day. In the first four months, Bob dropped 20 pounds and felt so much better he wondered why he hadn't tried such a program years earlier. The clincher that encouraged him to continue was a remark by a co-worker that he looked not only slimmer, but 10 years younger.

Avoid illness by learning to catch yourself before you're exhausted, chilled, overheated or in any state that's just this side of sickness. Taking time out to rest when you're in the middle of some difficult chore may be inconvenient, but if the rest break will save you from the discomfort of some illness, it'll be worthwhile.

In sum, start treating your body as if you were responsible for it, which you are. Don't forget, it's one of your most important assets. Keeping the rest of your possessions could depend on how well you keep your body.

And, by all means, cultivate the positive aspects of life and make a real effort to maintain an upbeat mental attitude. Boredom and depression have physical effects which make you more susceptible to illness. Anger and anxiety are other poisons, for which humor and courage are the antidotes. Staying interested, having your share of fun and enjoyment, and keeping yourself open to constructive emotions can be your best prescription for better health.

CONSUMER TIPS

Take Care of Yourself

Next to life itself, your health is your greatest asset. And, since the cost of health care has risen so dramatically in recent years, protecting this asset is an integral part of every family's financial security and well-being.

The key point here really is a combination of two points: Everyone needs to carry **enough** of the **right kind** of health insurance.

Some important checkpoints follow:

- ☐ See if you qualify—at work, through a professional association or by way of membership in any group—for a group health plan. This is the best way to save money on health insurance.
- ☐ Make health insurance benefits a matter of consideration when you contemplate a job change.
- ☐ Find a temporary policy to cover you if you face a period where you have no group coverage.
- ☐ Try to convert your group health insurance into an individual plan before you turn 65 so you will have supplemental coverage for Medicare and Medicaid.
- ☐ Discuss your needs with representatives of several insurance companies in case you are unable to obtain group coverage.
- ☐ Check to see whether there is an HMO or health service contractor in your area.
- ☐ Consider all the options and make your own decision about protecting your health.

Whether you participate in a group plan or purchase an individual policy, you'll want to know just what coverages you have so you can get maximum value for your health insurance dollars.

Check your group policy handbook, read your personal policy or consult your insurance representative to learn:

What coverages and payment limits exist under the basic hospital policy:

- ☐ Room and board
- ☐ Lab fees/X-rays
- ☐ Anesthesia
- ☐ Use of operating room
- ☐ Drugs/medications
- ☐ Ambulance service
- ☐ Other _____

Where major medical coverage takes effect.

What deductible must you pay on your own?

What coverages does your major medical provide?

- ☐ Hospital and surgical expenses
- ☐ Drugs/medicines
- ☐ Physicians' fees
- ☐ Nursing care
- ☐ Other _____

What is the maximum benefit?

Is it a lifetime benefit?

Is it adequate?

Is disability insurance a part of your coverage?

- ☐ Short term?
- ☐ Long term?

When Sickness Strikes at Your Savings

Dental insurance is a relatively new kind of coverage which could save you money. Do you have dental insurance?
- ☐ What does it cover?
- ☐ What are its limits?
- ☐ Is it available to you as an option under your group plan?
- ☐ Have you considered it?

Here are some things to check for when you examine a health insurance policy:
- ☐ What pre-existing conditions are excluded?
- ☐ What other exclusions are made?
- ☐ Is the policy guaranteed renewable?
- ☐ Is it noncancellable?
- ☐ What waiting period is in effect before payment begins?
- ☐ What is the benefit period?
- ☐ Have you elected, or should you elect, a cost-of-living rider on your disability coverage?
- ☐ Does the policy contain an incontestable clause?

MEDICAID
- ☐ If you are out of work or have a low income, you may wish to see if you qualify for Medicaid.
- ☐ Contact your local Social Security or welfare office for information about this government-sponsored benefit.

MEDICARE
- ☐ Consider a wraparound policy to supplement Medicare benefits.
- ☐ Know exactly what you're getting **before** you buy.

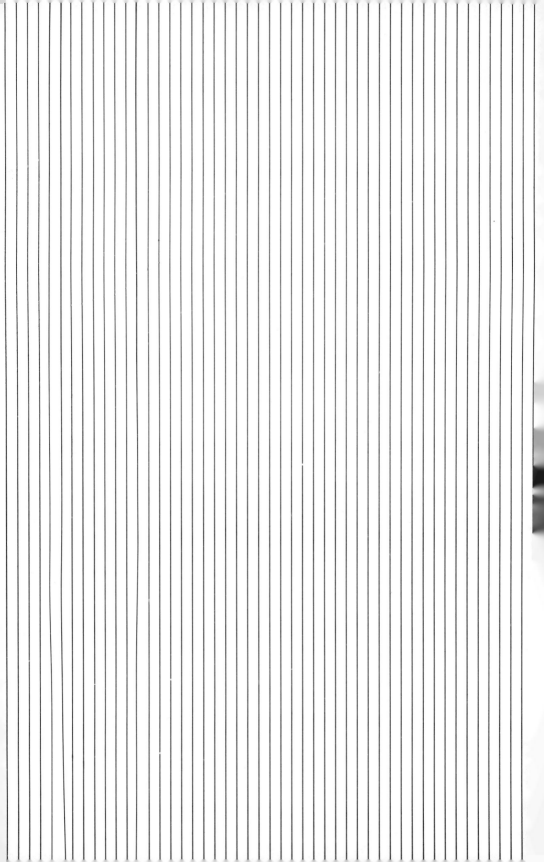

PART FOUR

Other Things You Should Know

CHAPTER 17

GUARDING AGAINST EVERYTHING ELSE

There are all the more commonplace ways of losing your possessions or your money, perils involving such things as your health, your car, your home, and your life, but there are also a number of lesser dangers, some of a highly specialized and rare nature. Sometimes you need to guard against disasters even from that quarter, especially if your circumstances put your earnings or your possessions in unusual jeopardy. For instance, in 1981, when Ronald Reagan was inaugurated, a baker named George Montilio of Quincy, Mass., was commissioned to bake for the President an inaugural cake. It was an elaborate affair, 7½ feet high and six feet long, all decorated with 50 eagles, 50 flags, portraits of Reagan and Vice President George Bush, and replicas of the Capitol Dome, the White House, and the Presidential Seal. To make certain that Montilio's time and investment did not turn to crumbs before the inaugural ball, the cake was insured for $6,000.

Your next lawn party may not be as elaborate as an inaugural ball, but if you're spending a lot of money for a catered outdoor wedding reception for your daughter you might want to get

Guarding Against Everything Else

insurance that it won't rain that day. Indeed, the insurance is available. It's available for a great range of things. You even can get insurance against the risk that your next vacation trip might be spoiled by rain. Groups sponsoring a "hole-in-one" contest at the country club can get insurance to guarantee the payoff just in case somebody actually sinks one.

Are you looking forward to the birth of a child? You can get insurance to help pay the extra costs if the offspring should be twins—or triplets, quadruplets or more. It's called "multiple birth" insurance.

You can get such special coverage just as you do your homeowners insurance, from your representative or agent. If your company doesn't offer the insurance, and your agent can't get it, they can refer you to a source that can, possibly an agent or broker specializing in "excess and surplus" lines of insurance.

The excess and surplus market is made up of companies that make a business of writing unusual, hard-to-get or one-of-a-kind insurance policies. Not all of them are small companies. Some of the largest regular insurers maintain excess and surplus divisions. One such group is Lloyd's of London, a collection of insurers that has the reputation for insuring anything that's insurable.

Lloyd's is where Jimmy Durante's nose was insured for $1 million. Various dancers and movie queens have insured their legs with Lloyd's. When stumped by the problem of defining the area of coverage requested by a belly dancer, a Lloyd's underwriter came up with a policy written for the dancer's entire body—"less head, arms, legs and chest."

A few other items that have been insured by Lloyd's:

- Elizabeth Taylor's Cartier diamond, given to her by Richard Burton, for $1.2 million.
- A grain of rice on which is engraved the portraits of the Duke of Edinburgh and the Queen of England, for $20 million.
- A 40-foot inflatable pink pig, used in concerts by the rock group Pink Floyd, for $1 million.
- The Human Fly, enroute across North America strapped to the outside of an airplane, for $1 million.

- A composer's ears.
- A whiskey distiller's nose.

Name it, and you can insure it, nearly always.

Aside from the excess and surplus market, many assorted risks are covered by "inland marine" insurance, written by large property/casualty insurance companies. Logically enough, inland marine includes "floater" policies, such as those covering personal effects, jewelry, furs, fine arts and the like. Although it sounds like a contradiction in terms, inland marine came by its name quite naturally, to designate insurance covering goods in transit, or articles that may be transported from one place to another—as well as bridges, tunnels and other transportation facilities. The name dates from the days when most interstate transport was carried by water. You might insure anything from a stamp collection to a shipment of statuary under an inland marine policy.

Here are some of the kinds of insurance that are outside the major categories of auto, home, life and health insurance, but which you as an individual might want to consider at some time. They are available or have been offered either in the regular or the excess and surplus markets. But remember, when you're deciding whether to buy any of these special coverages, weigh the premium cost against your own chance of loss. Sometimes your risk isn't equal to the cost.

Legal Expense Insurance

Group plans (sometimes referred to as "judicare") offered by several companies, will pay for legal services, from drawing up wills to defending against criminal charges, for a premium of about $100 a year. In some states, individual policies have been offered for a slightly higher premium. One plan features telephone consultation service for advice, legal problem-solving and routine chores, and also covers up to $10,000 in outside legal fees.

Travel Insurance

In addition to the one-shot life insurance policies commonly sold at airports, and accident policies often sold by mail, there are comprehensive travel policies that cover such additional eventualities as:

- Being stranded or losing your prepayment if an airline or tour operator defaults.
- Extra expenses if your luggage is misdirected or delayed.
- Extra expenses if a plane is hijacked, a cruise line goes bankrupt, or airline employees go on strike.
- Payment for your transportation to an emergency medical facility if a doctor orders it.
- Reimbursement if you cancel or interrupt a trip because of illness or other emergency.
- Compensation for the loss or theft of your belongings.

Investor Insurance

The policy offered by at least one company protects you against loss in case your brokerage house goes broke—an event said to have happened 135 times in a recent 10-year period. Losses could come from having a cash account with the broker, or from carrying stocks with the firm "in street name" without a certificate, or from maintaining a securities credit account. Most investors are protected already from such a calamity. The Security Investors Protection Corp., a federal agency, provides protection up to a half million dollars. This program is similar to the Federal Deposit Insurance Corp.'s coverage on your bank deposits. If you have more than a half million at stake you should consider this extra protection, but if you're only a small trader you're already protected and can save your money. One other form of investment insurance that you are not offered by the government is coverage for customers of commodity trading firms that go bankrupt. One private insurer offers that coverage.

Municipal Bond Insurance

This insurance guarantees payment of interest and principal on municipal bonds should the community or agency that issued them default. The coverage came out about the time that New York City and several other municipalities were having trouble meeting their bond guarantees.

Surgical Survival Insurance

This is a one-shot policy offered by a few companies that gives you extra life insurance protection if you should die within 30 days of a surgical operation—a risk usually considered uninsurable.

Child Care Insurance

If you're running a day-care center for small children in your home, you'd better have this policy, because homeowners insurance probably won't cover you (on account of the "business in your home" exclusion). Accidents can happen to the tots in your care. And liability for a serious accident can easily run into six figures or more.

Health Insurance for Pets

With the cost of veterinary care going up at about the same rate as medical care for humans, you might want to cover your pedigreed pooch or feline. Policies offered by several companies will cover ailments from ear infections to open-heart surgery, and injuries from such accidents as running into plate glass windows, falling off beds, and eating poison.

Mortgage Extra Expense Insurance

Available from at least one company, this insurance protects you in case your home is destroyed and your bank or other mortgage lender claims most of the proceeds, forcing you to refinance at higher interest rates in order to rebuild. The insurance covers the added cost to you of the new mortgage. It's a coverage you might want to consider if you have one of the older mortgages that carry extremely low interest rates.

Mugging Insurance

One company's policy, called "Violent Crime Victims Insurance," covers your property loss, lost wages, medical costs, rehabilitation expense, and payment for someone to do your housework, if you are assaulted, robbed or raped by a criminal. Another company will include payments for legal fees and even for your mental anguish. You're probably covered already for most of these costs in your regular insurance, so check first and don't spend your money needlessly.

Officers and Directors Liability

If you're a member of the town council or the board of directors of your local hospital you might well be in need of this special coverage. Few people even think about the dangers of being sued when on a public board until it's too late. Most corporations buy this coverage for their directors. And today people are suing not just the directors and officers of corporations but members of public boards as well. The supervisors in one Pennsylvania township were recently sued for trying to block a subdivision project which their voters had virtually mandated that they oppose.

Unemployment Insurance for Executives

If you make $25,000 a year or more, you might be interested in this coverage, which offers some improvements over ordinary government unemployment insurance. The insurance provides salary payments for up to two years, and the services of a professional job hunter or "outplacement" consultant. It covers dismissals due to layoffs, corporate takeovers, bankruptcies and personality clashes. It doesn't cover firings for illegal or unethical behavior, or practices prejudicial to the conduct of a business, such as coming to work drunk.

Other Things You Can Insure

If there's something you can't get insurance for, it's probably because nobody's thought of it yet. Some of the other coverages that have been thought of include insurance for:

- Adverse publicity
- Antique aircraft
- Auto racing spectator liability
- Contact lenses
- Customized vans
- Electric and hybrid vehicles
- Greenhouses
- Hang glider liability
- Home warranty against construction defects
- Houseboats
- Jet skis
- Mechanical breakdown (autos and mobile homes)
- Personal computers
- Pet mortality
- Saddle animal liability
- Sky sports
- Solar heating systems
- Stockbrokers errors and omissions

- Tax preparers errors and omissions
- Tuition refunds
- Waterbed liability

If the insurance isn't on the market in the form of an available policy, then you can probably arrange special coverage through your agent or broker, no matter how bizarre your needs. For example, there was the fence built by an artist named Christo. It was 24 miles long and 18 feet high. Officials of the California villages and towns through which the fence was to pass feared injuries in the crush of spectators and demanded that Christo obtain insurance. Finally, an insurer agreed to underwrite the fence.

The policy didn't cover the fence itself, which cost $2 million to put up and was to become worthless when it was taken down after two weeks. But it did provide Christo with several million dollars worth of liability coverage for a relatively small premium, and he went ahead with the fence as scheduled. Following his California success, Christo began looking for liability coverage for a new project: wrapping the islands of Biscayne Bay, Fla., in pink plastic.

CONSUMER TIPS

Your Insurance Portfolio

Now that you know you really can insure almost anything, you may want to assess your personal or business needs for any of these types of coverage in terms of protecting what's yours.

- ☐ Rain Insurance
- ☐ Multiple-Birth Insurance
- ☐ Personal Articles Floaters
- ☐ Legal Expense Insurance
- ☐ Travel Insurance
- ☐ Investor Insurance
- ☐ Municipal Bond Insurance
- ☐ Surgical Survival Insurance
- ☐ Child Care Insurance
- ☐ Health Insurance for Pets
- ☐ Mortgage Extra Expense Insurance
- ☐ Liability Insurance for Board Members
- ☐ Mugging Insurance

Guarding Against Everything Else

- ☐ Unemployment Insurance for Executives
- ☐ Other _____

To make a wise decision, it's essential to:
- ☐ Know the coverages you already have.
- ☐ Consider costs in relation to risk and ask yourself whether the product or event is worth insuring.
- ☐ Avoid duplication of coverages.

CHAPTER 18

WHEN YOU'RE THE VICTIM

Insurance can protect you from financial ruin if you should be responsible for someone else's injury. But what if you're the victim? Suppose you're the one who's hurt because of someone else's negligence. Then what protection do you have?

Now the shoe, as the saying goes, is on the other foot. And you have the same kind of protection, in most situations, as someone you might have injured.

Liability insurance is a two-way street. If you're legally liable for someone else's injury under conditions covered by your auto or homeowners policy, your insurer will pick up the tab, up to the policy limits. If you're the injured person, or it's your property that's damaged, in most cases you usually can file a claim under the other party's insurance policy. While that often may be an auto or homeowners policy, it can also be a commercial policy covering a businessperson or professional. For instance, if you slip and fall on a grease spot at work, breaking your coccyx, you'll probably be entitled to medical treatment and wage replacement payments under your employer's workers' compensation insurance. If a plastic surgeon's knife slips while you're having an operation, deadening a key nerve, causing you to lose your sense of smell, you can probably claim damages or sue, and be compensated under the doctor's medical malpractice insurance.

Or suppose you're attacked by a sofa bed, as one woman was. She'd put the linen away in a compartment in the bottom of the sofa bed, and was sitting in front of it reading an old letter she'd found there. Suddenly a spring broke loose and struck her in the right eye, blinding her on that side. She sued the manufacturer and won a substantial judgment.

In most cases involving a manufactured product, the claim or award would be paid by an insurance company under product liability coverage—or a clause in a comprehensive liability policy—protecting manufacturers and distributors. This insurance covers against claims by people who are injured or who have losses as a result of using their products.

A case where the plaintiff wasn't injured, but was compensated for a sizable financial loss, was that of an Eldorado Springs, Mo., dairy farmer to whom a jury awarded more than $1.2 million in damages from the makers of an electric milking machine. It seems the machine delivered shocks to the farmer's cows while it was milking them. The cows protested by drying up, to the farmer's great bafflement and financial loss. The cause wasn't discovered for nearly a year. During that time milk production fell off drastically and more than 90 of the farmer's 130 cows died from infections and udder problems. Light dawned on the farmer when he read an article in a dairy magazine about the effects of stray voltage on cows.

Professionals are covered by policies under a wide variety of labels. Basically, you'll find you're protected against their mistakes under one of two forms of professional liability insurance:

- Malpractice liability insurance, which applies to physicians, surgeons, dentists, nurses and other professionals whose services involve touching the body;
- Errors and omissions insurance, for lawyers, architects, insurance agents and other professionals whose services do not usually involve body contact.

These are people whose services are expected to meet consistently high standards, and whose mistakes can affect not only their pocketbooks but their reputations. Hospitals, which provide both services and products—such as medicine and food—usually carry policies combining malpractice and comprehensive general liability coverages. If your coat is stolen or damaged by fire while at the cleaner's, chances are you'll collect under a liability policy known as bailee's customer insurance. (A bailee is anyone who has rightful possession of someone else's property.) Bailee's customer insurance usually is carried by laundries, department stores, repair shops and other establishments that take custody of customers' goods for repairs, renovations or other services.

There are many other examples of special purpose liability coverages, but that's probably enough to give you the picture. Nearly anyone, in any line of business, can find an insurer to issue a liability policy to fit his particular exposures or risks.

How to File a Claim

To file a claim, all you ordinarily have to do is call or notify in writing the person or organization responsible for your loss—or its insurance company. The insurance company will respond. Probably a claims representative will contact you and an adjuster will be sent or someone will interview you to obtain the necessary details. If the case is complicated, or if there's a disagreement over the facts or the amount of settlement, you may want to have a lawyer represent you. That doesn't necessarily mean going to court. In most cases, the lawyer will attempt to negotiate a settlement offer with the insurance company. This is cheaper and quicker for you than going to court.

In all cases, you should consider seriously the company's offer, which ordinarily is based on its realistic estimate of the "value" of the case, based on all the facts available.

A Houston, Texas, woman had just put her house on the market in the expectation of a sizable profit from its sale, when her young son, playing with matches, set the house afire, causing extensive damage. The insurance company adjuster, after a thorough examination, made her an offer great enough to repair the damage. Her hopes of a profitable sale upset, the woman brought in outside contractors, some of whom were her friends, to give additional estimates. With these higher estimates in hand, the homeowner refused the company's settlement offer, and the case went to arbitration. The arbitrators, whose decision was binding, decided that the additional estimates were inflated, and ordered the insurer to pay her less than the company's original offer.

In accepting an insurance payment for serious personal injury, you might want to consider what is called a "structured" settlement. This is where the insurance company, instead of paying you a lump sum, pays you a stipulated monthly payment over a period of years or for life.

When an insurance company offers a settlement you're not satisfied with, you do have the option to sue. In such a case, your chance of collecting depends on your ability to prove your case. But you should keep in mind that if you go to court, you could lose the case or come out with less than originally offered. Even if you win you'll have to give your lawyer a hefty cut, usually one-third, of any court award, depending on the agreement made.

What if the person or firm that causes you harm doesn't have liability insurance?

If you're injured in an auto accident and you have uninsured motorists coverage, you should be able to collect for your losses from your own insurance company. Your auto collision insurance will cover damage to your car. In most other cases, you'd probably have to sue if you couldn't talk the other party into paying for your damages voluntarily. But the amount you could recover would be limited by the other party's ability to pay (lawyers frequently refer to indigent defendants as "judgment proof," saying "you can't get blood out of a turnip").

Workers' Compensation

If you get hurt on the job, or if conditions at work damage your health, chances are about nine to one you'll be eligible for workers' compensation benefits. That's roughly the proportion of U.S. employees who are covered by workers' compensation insurance plans set up by laws enacted in every state and jurisdiction of the United States.

In most cases, benefits are paid by private insurance companies under contract with employers in accordance with requirements of the state laws. In a few states, the benefits are paid from state insurance funds—in effect, state-owned insurance companies.

Over the years, the types of injuries and illnesses covered by workers' compensation have been expanded greatly, and compensation boards have given increasingly liberal interpretations of eligibility. In some cases even ailments such as heart attacks have been compensated on the grounds they were caused by occupational stress.

Sometimes workers' compensation claims can border on the bizarre. Seven employees of a company received workers' compensation after an unstable fellow employee who was discharged apparently decided to do some firing of his own, came back the next day with an automatic weapon, and shot them.

Sometimes, the circumstances create unusual problems. One case involved a young worker who was assisting in the blasting of a well. After placing the blasting caps in the well, he detonated the charge by hooking up the lead wires to his truck battery. The blast blew the wires out of the well and onto high tension power lines overhead. Unfortunately he didn't let go of his end in time and was electrocuted. After the payment of death benefits to his parents, it was discovered that the man had fathered an illegitimate child, and an Alabama court got the task of deciding whether the child was eligible for a dependent benefit.

Workers' compensation is a type of no-fault insurance (one of the earliest, dating back to the turn of the century). This means that the employer assumes responsibility for the costs of work-related injuries and illnesses, regardless of fault—either his or yours. In return for this blanket protection, you as the employee give up the right to sue the employer in most cases (courts have allowed some exceptions to the rule). Rather than bear the costs of each case directly, most employers elect to protect themselves by buying an insurance policy covering all employee claims.

This arrangement is not a cheap "out" for an employer. On the contrary, insurance premium rates are usually based on a company's safety record, giving employers a powerful incentive to make the workplace as safe as possible. When losses, and rates, get too high, they can add significantly to the cost of doing business. In addition, many insurers will not accept a company as a customer until its plant passes the insurer's safety inspection. So effective are these incentives that workers' compensation insurance has been credited with being the main force behind the U.S. industrial safety movement.

Benefit levels are set by the state legislatures, and vary from state to state. Generally, workers' compensation pays for:

- Hospital, doctor and other medical expenses;
- Weekly payments for a percentage of lost income, up to a maximum amount, while you are disabled or if you return to work with decreased earning power;
- Rehabilitation therapy to enable you to return to your old job or train for a new one;
- Survivor benefits to your dependents if you should die because of a work-related accident or disease. Death benefits usually are paid to the spouse until remarriage and to the children up to a certain age, and include a percentage of former weekly wages as well as a burial allowance.

Benefits are based on various formulas set up by the states. But usually they are based on the employee's weekly salary and the average weekly wage for all workers in the state. (For example, disability payments might be set at two-thirds of your pay, up to 120 percent of the general average wage—or some other specified numbers.) Some states put limits on the number of weeks you can collect for certain types of injuries. Benefit levels in most states have been raised in recent years.

As of 1982, workers' compensation insurance was compulsory in all but three states—New Jersey, South Carolina and Texas, where it is optional for employers. But even in these states, most businesses carry it in order to limit their exposure to negligence suits. Or, to put it another way, to substitute a manageable known cost (the insurance premium) for the unpredictable cost of a liability suit that could have the potential to cripple them financially or put them out of business.

State workers' compensation laws cover workers in nearly every kind of job. Some states exempt a few occupations—most commonly, agricultural workers, domestic help, casual laborers, public servants and employees of non-profit groups or businesses that employ only a handful of people (fewer than five in some states; three or less in others).

Some workers not covered by the state laws are protected under other statutes. These include employees of the federal government (Federal Employees' Compensation Act), railroad workers (Federal Employers' Liability Act) and maritime workers other than seamen (Longshoremen and Harbor Workers' Act). Seamen are covered under the Merchant Marine Act.

Other Help for the Victim

Although you're covered by other people's liability coverages if they, their cars or their products cause you to be injured, there are other kinds of liability insurance that could help you in unusual situations, such as volcanic eruptions and nuclear accidents. Your homeowners policy covers your property if it should burn because of a nuclear incident; other kinds of losses caused by nuclear reactors, radiation or radioactive contamination are excluded under most homeowners and other personal policies. But that doesn't mean you're left unprotected if you should become the victim of some rare nuclear incident.

Three special insurance pools, combining the resources of many property and liability insurance companies—and backed by the federal government in case of a catastrophic accident—stand ready to pay claims by the public for personal injuries or property damage caused by a nuclear incident. Under the terms of the Price-Anderson Act, the limit of liability for each nuclear reactor is $560 million for any one accident. The nuclear insurance offered by the pools is purchased both by the nuclear power operators and any firms that supply their materials and equipment, or which transport nuclear materials.

The first real test of the nuclear insurance system came on March 28, 1979, at the Three Mile Island nuclear generating station in Pennsylvania. The malfunction which shut down the plant caused no deaths or injuries, and no off-site property damage. But the state's governor recommended the evacuation of some people from the area, and the insurance pools agreed to pay for the costs of evacuation.

Altogether, the pools wound up paying evacuation claims totaling some $1,213,000 to 3,160 claimants, plus lost wage claims of about $90,000 to 615 individuals. Insurance also paid for $300 million in losses from damage to the facility itself.

Another kind of catastrophe that caused massive damage and many casualties came on May 18, 1980, when Mount St. Helens in Washington State began a series of volcanic blasts that devastated millions of dollars worth of property and forests and covered thousands of acres with dust and ashes.

The incident created some confusion about insurance coverage. The problem was that traditional homeowners policies specifically excluded "volcanic eruptions," but the simplified policies of the late 1970s did not have that exclusion. Since both types of policy cover damage caused by "explosions," most insurers gave a liberal interpretation to the coverage and paid the vast majority of property damage claims in the blast area under that clause. To prevent future confusion over coverage, the state insurance commissioner and the insurance companies later agreed that homeowners and fire insurance policies that cover volcanic eruptions would spell out the coverage more explicitly.

What these examples illustrate is the fact that if you are hurt or have a loss from property damage through someone else's fault—or even through natural or man-made disaster—there usually is insurance of some kind that will help you.

CONSUMER TIPS

Are You the Victim?

Remember that liability insurance works two ways:

- ☐ It pays when you are responsible for injuries to other people or damage to their property.
- ☐ It pays when you're the victim of injury or when it's your property that's damaged and you file a claim under the other party's insurance policy.

A special form of liability insurance—product liability coverage—will pay your claim should you be injured while using a defective manufactured product, such as a household appliance.

Professionals purchase one of these kinds of liability coverage so you're protected financially against their mistakes:

- ☐ **Malpractice liability insurance**, which applies to physicians, surgeons, dentists, nurses and other professionals whose services involve touching the body;
- ☐ **Errors and omissions insurance**, for lawyers, architects, insurance agents and other professionals whose services do not involve bodily contact.

Most businesses carry a comprehensive general liability policy. (The "bailee's customer insurance" which pays you if your coat is stolen from the cleaner is one example.)

HERE'S HOW TO FILE A CLAIM IF YOU'RE THE VICTIM:
- ☐ Phone or write the organization responsible for your loss or its insurance company.
- ☐ The insurance company probably will have a claims adjuster interview you to obtain the necessary details and then to offer a settlement.
- ☐ If you are dissatisfied with the settlement offered you by the company, you may want to have an attorney represent you as a negotiator.
- ☐ In the event that a satisfactory settlement cannot be worked out, you have the option to sue. Keep in mind, however, that you could (a) lose the case or (b) come out with less than the original settlement offer, even if you win.

HERE'S WHAT TO DO IF THE OTHER PARTY HAS NO INSURANCE:
- ☐ For auto-related injuries, collect for your losses under your own uninsured motorists coverage.
- ☐ For damage to your car, collect from your own collision coverage.
- ☐ For other situations ask the responsible party to pay voluntarily, or sue the other party.

Workers' compensation benefits will pay if you're hurt on the job. Check with the person at your place of work who handles insurance coverages if you have questions.

PART FIVE

Protecting Your Nest Egg

CHAPTER 19

WHEN RECESSION COMES

Next to the health and lives of you and your loved ones, your most precious possession is your livelihood. After the Great Depression of the 1930s, Americans believed the controls the government had imposed on our economy would prevent another depression from ever occurring. Nevertheless recessions, or periodic downturns in the economy, do come, often with vexing frequency and regularity, and some of those slumps are so severe they bring back grim echoes of the Great Depression itself. When 10 percent of the nation's work force is unemployed, as has been the case recently, not only are those millions of laid-off workers affected but millions more find their own jobs threatened and their salaries and investments eroded. Yet there are ways to guard against the worst of the impact.

Guard Your Job Skills—Whether you are a production worker or a manager, the one thing in this world that you can sell is that collection of experience, training and abilities known as job skills. You not only want to enhance those skills, you need to ensure that they do not lose the market value they have acquired thus far in your career. The first thing you must do is keep a regular, watchful eye on the state of your particular industry and, in particular, the health and future prospects of your company. No one at the Pennsylvania Railroad ever thought that the "Standard Railroad of the World" would ever go out of business. But, when it came in 1970, the Penn Central crash was the biggest bankruptcy in history. Ten years ago, few employees of the Big Three auto companies ever dreamed that their jobs would be in jeopardy because of a rising flood of European and Japanese imports. But today tens of thousands of America's auto workers are idle, and at least one big U.S. car maker has been put through a wrenching financial squeeze. And the loss for many workers is permanent. Many have had to move to other parts of the country and seek jobs in other industries, often learning new trades at much lower wages. So it makes sense to keep an eye on the direction your company is taking and find a new corporate home if the future health of your firm appears doubtful. The same goes for your industry. If you are in some computer-age equivalent of the buggywhip business, it might make sense to learn a new trade. And practice and broaden all your skills. Don't stop educating yourself. Don't stop learning everything you can about new technology. And don't pass up the chance to learn other trades that you might switch to if things get tough in your own line of work. It's as important to have as modern and diverse a selection of work skills as it is to have a strong investment portfolio. For instance, if you're a typewriter repairman you'd better start learning about computers and word processors, or technology is going to make you outdated in the employment market.

Be Careful With Debt—It's easy and quite appealing to buy now and pay later. When times are flush it's a simple matter of taking on a loan, buying at today's prices and paying for it in tomorrow's inflated dollars, when your income will probably be 10 or 20 percent higher. In theory this works quite well. In practice it has worked quite well, too, for many people. But not lately. When a recession comes and inflation slows down— and that's what's happening today—you could well be stuck with a lot of debt and interest payments at a time when your income is sitting still or even falling. And if you bought and borrowed at a time when a lot of other people were doing the same thing you may well be paying a rate of interest that is downright backbreaking. So, be careful.

When Out of Work, Set Up Priorities—If you should land in the street, make sure you use what income you have in the best possible way. First, make sure you're collecting all the money you are due. Make certain you get your proper severance. And whatever union benefits you are owed. And don't forget your unemployment insurance. That done, set up a list of priorities for paying bills. Of course food is a first, as are heat and lights. But then look at what money you have left and budget it so that your most important possessions are protected from foreclosure or loss. Real estate taxes should be paid. Otherwise the sheriff could auction off your property. Insurance premiums, certainly those on the coverages that are most essential, such as your house, personal possessions and car, should not be allowed to lapse. And if you need to borrow money, look carefully at interest rates and choose only the source of money that is cheapest. One good place to turn for emergency funds is your life insurance policy, where interest rates for loans often are lower than at the banks.

Watch Out for Inflation—Regardless of how pressing a recession can be, never forget that inflation is probably right around the corner, and plan accordingly. Work out a strategy for financial and career management that takes inflation into account. Never forget, the economy, with all its uncertainties and whims, whether it rides on the vehicle of recession or inflation, portends the most constant and pressing threat of all to the growth of your investments, to the value of your dearest and most costly possessions, and to the continuance of your livelihood itself.

CONSUMER TIPS

On the Job

- ☐ Enhance and broaden your skills. Learn and practice everything you can in the realm of new technology.
- ☐ Keep a close watch on the health and future prospects of your industry and, in particular, your company.
- ☐ Research the possibility of a new job if the future appears doubtful.
- ☐ Be careful with debt.

Out of Work?

- ☐ Set priorities.
- ☐ Use your limited income in the best possible way.
- ☐ Make sure you're collecting all the money due you.
- ☐ Set up a list of priorities for paying bills. After food, heat, and electricity, be sure to keep real estate taxes and insurance premiums firmly in mind.
- ☐ Should you need to borrow money, choose the source with the lowest interest rate. Your life insurance policy is sometimes a good source of emergency funds.

Any Time

- ☐ Beware inflation!
- ☐ Work out a strategy for financial and career management that takes inflation into account.

CHAPTER 20

KEEPING YOUR NEST EGG FROM BREAKING

Of all the things that are yours, one of the most crucial to your well-being is your nest egg, your savings for the future. Whether in the form of a passbook savings account or a condominium on the Gulf Coast, the extra capital that you've salted away is your security for old age or an emergency. Yet, when it comes to nest eggs, inflation is like a snake in a bird's nest. If you don't manage your nest egg right, inflation will eat it up.

For instance, many people with a few dollars left from their paychecks are willing to put their money into a bank savings account, earn passbook interest, and leave the reading of books, financial news, corporate annual reports and prospectuses to others. But inflation makes nonsense of such a passive approach. If your passbook account pays 5 percent interest each year, and the annual inflation rate is around 10 percent, it's plain you're losing money. The value of your capital is shrinking. The basic idea in any plan for building a nest egg is to have your money grow faster than the rate of inflation, after you've paid the taxes on the growth. There are various ways of accomplishing this, but all of them take a little thought and study.

For many people, the best and simplest way to get a relatively high rate of return and keep their money safe is by investing in certificates of deposit, or CDs. They and similar higher-interest instruments are available from banks or thrift institutions and guaranteed by the federal government or one of its agencies, a feature that provides a degree of safety that many other forms of investment cannot provide.

Nevertheless, there are many other ways to invest that you should consider as well. Alternatives include common and preferred stocks, convertible securities, corporate bonds, government bonds, issues sold by government agencies, tax-exempt issues, Treasury bills, mutual funds, money market accounts, options, commodities, precious metals, real estate and collectibles.

Savings Deposits

You should have a passbook savings account, or something equally accessible, so that you can get to an emergency source for funds quickly if the situation demands it. But the rest of your savings are better invested in something that pays you more, a certificate of deposit for instance, or one of the other basic forms of investment listed above. Usually, passbook savings accounts pay far less than the rate of inflation. In many cases you are better off merging your savings with your checking deposits and opening a "Now" account, a checking service that pays interest on the balance. Most "Now" accounts require a minimum balance that is well above the minimum required for most free checking accounts. But if you merge your passbook and checking accounts your balance will probably come close to exceeding that minimum. The only problem with the "Now" account is that it usually pays less than the inflation rate, so—just like the passbook account—it's good only for emergency cash savings, not for the bulk of your nest egg.

Money Market Deposit Account

There's a far better way to lay aside instant cash for an emergency than putting it in a passbook or a Now account. Take a look at one of the new money market accounts that banks are now offering. In an effort to compete with money market funds offered by insurance companies and brokerage houses, the banking industry recently began offering savers its own version of this popular account.

WHERE AMERICANS PUT THEIR SAVINGS

	Percentage of Total Savings
Savings accounts, time deposits	28.8%
Corporate stock	22.3
Life insurance, pension reserves	21.6
Demand deposits, currency	6.4
U.S. government securities	5.1
Money market funds	3.9
Corporate bonds, open-market paper	2.3
State, local securities	1.9
U.S. savings bonds	1.4
Other investments	6.2

Source: Federal Reserve Board.

Whether it's called "Liquid Investment Account," "Money Management Account," "Insured Money Market Account," or any of the variations of the theme, the basic concept is the same from bank to bank. For a minimum of $2,500, a depositor can open an account that will pay interest at money market rates, generally higher than those paid to regular savings accounts. In addition, all funds up to a maximum of $100,000 are insured by the Federal Deposit Insurance Corporation—a feature not offered by money market funds.

Most banks allow you the right to withdraw or make deposits in this account without penalty, and also offer limited check-writing privileges and the right to transfer money into another account. In the event that your balance falls below the $2,500 limit, the higher rate of interest would not be paid.

It is worth noting that the interest rate can vary from one period to another and that the banks have the right to modify or change the terms of the account at any time.

Common Stocks

More than 30 million Americans own common stock, meaning that there is one shareowner in every four U.S. households.

That means that after various bank savings programs, common stocks are the most popular form of investment. Common stocks are suitable for those who want to take an active part in building their nest egg.

If you decide to invest your money in common stocks, you should remember that in today's fast and complex world you no longer can buy and forget. The stocks you are considering should be researched before purchase; the broker you select should be reliable; annual reports and quarterly statements should be read, dividends monitored, and all issues watched with constant vigilance—whether you own a single stock or a large portfolio. And you must develop a clear program of investment. Your money is too important to be mislaid.

Most people have a favorite story about an Uncle Harry or Aunt Sarah who bought stock in a now-defunct company, which they held until they died, and the worthless certificates, found in a trunk, were given to the children to play with. Occasionally there is an opposite case like that of a church which was left some shares in an apparently moribund German oil company. The church became wealthy when the company made an unexpected oil find, and wealthier still when the world petroleum price quadrupled. Then, too, there are rare cases where stocks of seemingly defunct companies can be traced to currently thriving concerns which absorbed the old companies.

Choosing which stocks to buy is clearly a monumental task. There are more than 1,500 common stocks listed on the New York Stock Exchange, 850 on the American Stock Exchange and more than 7,500 on regional exchanges and in the Over-the-Counter (OTC) market of stocks that are traded privately between brokers but not listed on any exchange. The selection ranges from conservative to speculative, from stocks that have paid dividends for many years to those that have never yielded anything. The industries included cover the spectrum of the American economy from aerospace and autos to transportation and utilities. Most of the nation's major companies have their shares listed on the New York Stock Exchange. Other companies are found on the American Stock Exchange or in the sprawling OTC market. Your stock broker can carry out your buy and sell orders in any of these markets.

Your decision on what to buy depends on who you are, and on your abilities and needs. The right choice hangs on your age and family status (single, married with no children, married with dependent children, or married with children who have grown and left home). You must take into account whether you are young and building for your future or moving toward retirement. You have to consider your job and its fringe benefits. The stock selection process will differ depending on whether you are at one end of this scale or somewhere in the middle. For example, if you are just starting your career, dividends should be of secondary importance. The big thing is growth. But as you move toward retirement, preservation of capital and dividends becomes more important.

The current tax structure should be considered as well. The existing tax law states that only the first $100 of dividends may be excluded from your income tax, or $200 for those filing a joint return. So, for a young couple to load up on dividend-paying issues that show little or no growth could prove costly.

Historically, people have invested in common stocks because they have outperformed other investments over an extended period. Over the past half century common stocks have shown a 9 percent average annual gain, compared with about 4 percent for fixed-income issues. But in recent years, the stock market as a whole has not kept up with inflation. This has been due partly to the surge in interest rates for other forms of investment. When interest rates have fallen, stocks have usually climbed.

Selectivity is the key to any stock program. The market can remain flat while many investors make money buying and selling and taking profits on the quick fluctuations of selected stocks.

If you should invest in stocks, your best strategy is to spread your money over a carefully balanced cross-section of the three basic types of equities:

- Those stressing growth of capital.
- Those stressing income, or dividends.
- Those seeking to provide a mix of both dividends and capital growth.

Growth Stocks—Growth companies are those that enjoy rapid increases in sales and earnings from year to year. Frequently, managements of such companies pay only minuscule dividends or none at all and reinvest earnings in new plants and equipment, the expansion of markets, and the development of new products. In recent years such industries as medical technology, computers, drugs, fast food services, health care facilities, automation, and advanced technology have enjoyed above-average growth.

When you purchase a growth stock, it's worth remembering that the issue will more than likely be volatile, often fluctuating in price. At the same time, this type of stock continues to offer the best hope for capital growth over a period of years. A true growth company will have achieved a record of consistently rising sales and earnings over an extended period.

But beware. The greatest danger with growth stocks is the strong chance the companies' earnings may decline. The market usually reacts harshly to disappointment. Sharp price drops are frequent when a growth company announces lagging results. The true growth company must be able to maintain its record of earnings growth through rising, flat and declining economic trends. Thus, the investor in a growth stock—or a portfolio of such stocks—must watch each industry and company with care and be alert to any slowdown. Knowing when to sell a stock is perhaps even more difficult than knowing when and what to buy. But hanging on to a stock that is headed for the cellar—or the graveyard—is not the formula for financial success. At times, you must be prepared to "cut your losses"—that is, sell at a small loss to avoid a larger one later.

Income Stocks—The alternative to growth stocks is income stocks. These are defined as issues with a solid history of dividend payments, and the ability to increase these payouts regularly. Also needed is a record of rising earnings strong enough to maintain the policy of dividend increases.

Over the years, the utility industry has been the best group for this kind of investment. Utilities generally have benefited from fairly constant rates of return assured by state regulators, and a lack of competition. Although there are scores of attractive utility issues available that meet income-producing requirements, some do have drawbacks to the investor. Some electric utilities have heavy commitments to nuclear power that are proving costly. They and others are facing environmental problems that could cut into their future earnings. These factors should be weighed when selecting a particular issue.

One recent development has made the utility group particularly attractive to investors who want income and growth. Under the Economic Recovery Tax Act of 1981, investors in utilities now receive a significant benefit. If the utility you invest in meets certain conditions laid down by the government, you may exclude from your taxable income up to $750 a year, or $1,500 on a joint return, from your dividend if you participate in the company's stock dividend reinvestment plan. Taxes on the excluded amounts are deferred until you sell the shares, when they are taxed at the rate for capital gains. Such plans, which many companies now offer their shareholders, permit investors to take additional shares of stock rather than cash when dividends are paid. If you have no immediate need for income but want a dividend-bearing stock, you should consider this plan. But remember, this really is no different from a long-term return on your investment, and if that's what you want, you'll probably make more money with growth stocks.

Total Return Stocks—Growth and income stocks have been joined in recent years by the stock for total return. With any investment—whether a single issue or a portfolio—the investor usually must make a trade-off between safety of principal (and perhaps a loss in "real" dollars due to inflation) and acceptance of greater risks in order to achieve capital appreciation.

The total return approach is designed to provide high current income and preservation of capital while at the same time allowing the opportunity for capital growth. The investment may be common stocks, convertible securities, fixed income issues or a combination. The rate of return is figured as a combination of yield and capital appreciation. For example, ABC company may project a 12 percent price appreciation; XYZ company may offer a 12 percent dividend yield. But for total return, you'd be better off with LMN company, which offers you a 10 percent yield plus 5 percent price gain.

Preferred Stocks—Falling somewhere between a bond and common stock is the preferred stock. It pays a fixed return or dividend, although the dividend is not guaranteed. But preferred stock gets priority, and the dividend payment would have to be made on the preferred stock before the common stock could receive a dividend. Price movements for preferred stocks usually are in line with corporate bonds, rather than with common stocks.

Protecting Your Portfolio

Regardless of what type portfolio you select—and it may be a combination of types—it is vital that you do your homework and bookkeeping. Just to put the stock portfolio away for a rainy day could prove extremely costly. When to sell a stock is at least as important as what to buy—and you can't make intelligent selling decisions unless you follow both the market and your own particular stocks on a weekly, if not daily, basis.

You can choose from a number of good market letters to help you with your decisions. And market advisers and stock information and counselling services are available as well. In addition, there are books on the market that will provide detailed guidance about various types of investments and how to formulate the right investment strategies.

No matter what your investment or how limited your financial knowledge, you should take certain steps to ensure that your hard-earned money goes on earning more money. If your investment dollars are in common stocks or mutual funds, you should always make sure that you examine all the information available about the company. This is done through annual reports; all corporations that are publicly held must issue them. The annual report is sent to shareholders by the company or mutual fund organization. Most companies and the majority of mutual funds also issue quarterly reports. Some mutual funds make only semiannual mailings to shareholders. That should be the minimum expectation from any stock or fund you own. Most firms also have shareowner relations departments that will answer any questions and provide you with any information you may need.

Make a regular habit of scanning your local newspaper's financial section. Big-city papers have more extensive coverage, but virtually all Sunday newspapers have financial sections to provide you with some information. For those interested in more detailed information there is The Wall Street Journal, the nation's leading financial daily. On a weekly basis, Barron's can be most helpful. In addition, there are scores of financial magazines. The choice obviously depends on how much reading you wish to do and how extensive your portfolio holdings are.

For those who make their own investment decisions, there are many investment advisory services available. Among the largest are Standard & Poor's Outlook, Moody's Investment Service and Value Line Investment Service. These publications are sold on a subscription basis, and the cost is tax-deductible. Often, they can be found in libraries. There are also monthly, biweekly or weekly market letters as well that can be useful for active participants in the stock market.

Investment counselors provide another source of advice and information. This method may suit you if you have a large portfolio (many investment counselors won't handle accounts of under $100,000) and if you lack the time to give it proper supervision and care. Investment counselors operate on a fee, depending upon the size of the portfolio. Many investment counselors also provide a variety of other services, including record keeping and handling of dividends, and some will even make purchase and sale decisions for you—provided you give them a power of attorney.

Many of the services provided by an investment counselor are available as well at savings or commercial banks. Any fee structure depends on the size of your portfolio.

Finally, there is your local registered representative, or stock broker, working for a brokerage house. In most cases, he or she is your link with your investments. Most brokerage houses also have research departments that follow the stock market and many individual issues. Whether you are an active trader or just use the broker's services occasionally, information about your holdings should be readily available from this source.

You might be nervous about making stock selections on your own, and yet not want to make use of a mutual fund. If you still wish to take an active part in the stock market and the selection process, there are some alternatives. One is the investment club.

The club is a group whose members get together regularly—normally once a month—to pool a designated amount of money, usually between $25 and $100 per person, and then discuss various stocks. After some discussion, the group agrees on one of the issues and it is purchased as soon as possible after the meeting.

Investment clubs can be formed by friends, neighbors, groups of fellow workers, members of baby-sitting pools, bowling clubs, church groups, social organizations, or in fact just about any collection of people. The primary requirement is that they share an interest in the market, be willing to listen to each other and recognize that investing is a risk. The ideal investment club should have about 15 members, although there are some as small as 10 and others with as many as 25 or more members. Results vary from one club to another, just as they do with mutual funds. Some clubs do extremely well, and many break up after a time. But a club that has been operating for several years probably is functioning to the satisfaction of its members, and it is sometimes possible to join or buy into an existing club. At a minimum, the clubs provide a way of building your nest egg in a fashion that's interesting and enjoyable.

Another possible approach is to give discretionary power to a broker or money manager to buy and sell for your account. If you take this route, you had better know this individual well and have complete confidence in him—and also make regular appraisals of his performance for you.

Bonds for Income

If you are leery about buying common stock on your own or joining an investment club or accumulating shares of a mutual fund, there is another attractive vehicle for you: bonds.

Bonds are instruments of credit. Or to put it more simply—bonds are a corporate or government IOU. The organization issuing the bond promises to pay you a certain amount of money on a certain date. In exchange for that promise, you lend the issuer the money. In order to make this transaction attractive, the company or government will also pay you a fixed amount of money—interest—twice a year. When the bond becomes due, or matures, you will receive the face value, or the original amount lent. Bonds, like stocks, are traded on the open market, and there could also be a possibility of capital gain if you bought a bond at a discount and later cashed it in at face value or sold it at a higher price.

Many different types of bonds are available. Organizations issuing them include federal, state and local governments as well as corporations.

The safest and most secure bonds are those issued by the United States government. They include three-month Treasury bills and longer-term U.S. Treasury bonds, as well as obligations issued by various federal agencies—Government National Mortgage Association ("Ginnie Maes"), Federal National Mortgage Association ("Fannie Maes"), Federal Home Loan Bank Board, Tennessee Valley Authority (TVA) and others. Backed by the full faith and credit of the U.S. government, they enjoy the highest credit rating available and no issue has ever defaulted or missed an interest payment. A ready market exists for these issues. The initial purchase price usually is $10,000 or higher, but some $5,000 issues are now available. Taxes must be paid on interest from agency obligations, while Treasury bills are free of state and local taxes, but taxed by the federal government.

Among alternatives to bonds issued by the U.S. government and its various agencies, and probably as attractive if not more so, are municipal bonds. Free of federal, state and local taxes in the state of issuance as well as some other states, "municipals" as a category represent the obligations of state and local governments and often are issued to provide for such things as public housing, college dormitories, or bridges and tunnels. They are also issued to cover general obligations of the state or municipality.

Municipals come in two forms—short-term issues or notes which are issued by state or local governments to raise money in anticipation of revenues, and long-terms bonds which are sold to finance various projects.

As with federal government securities, it takes at least $5,000 to purchase a single municipal bond, while notes are usually available in the minimum denomination of $25,000. Since the number of local governments offering bonds is large, it becomes almost impossible for the average person to acquire a representative number, and to keep track of all the available issues. Nevertheless, individual investors own a hefty share of municipal bonds, as the accompanying table shows.

HOLDINGS OF STATE AND LOCAL GOVERNMENT SECURITIES

	Amounts Outstanding* (Billions)
Commercial Banks	$151.1
Individuals	62.9
Corporations	3.4
Life Insurance Companies	6.7
Other Insurance Companies	83.5
Mutual Savings Banks	2.7
Others	15.1

*Estimates as of Jan. 1, 1981.
Source: The Public Securities Association.

As a result of the growing demand by investors for a greater opportunity to participate in the benefits available from the ownership of both U.S. government and municipal securities, mutual funds and investment trusts specializing in both types of securities have become increasingly popular.

In the case of the investment trusts, the managing trustee selects a diversified portfolio of either government or municipal securities, collects and distributes all interest income to the trust

Tax Exempt/Taxable Yield
Equivalents
(Individual income brackets—thousands of dollars)

Single Return ($000)		$18.2 to $23.5		$23.5 to $28.8		$28.8 to $34.1	$34.1 to $41.5		over $41.5
Joint Return ($000)		$24.6 to $29.9	$29.9 to $35.2		$35.2 to $45.8		$45.8 to $60.0	$60.0 to $85.6	over $85.6
% Tax Bracket		29%	31%	33%	35%	39%	40%	44%	49%	50%

Tax Exempt Yields (%)									
6.0	8.5	8.7	9.0	9.2	9.8	10.0	10.7	11.8	12.0
7.0	9.9	10.1	10.4	10.8	11.5	11.7	12.5	13.7	14.0
7.5	10.6	10.9	11.2	11.5	12.3	12.5	13.4	14.7	15.0
8.0	11.3	11.6	11.9	12.3	13.1	13.3	14.3	15.7	16.0
8.5	12.0	12.3	12.7	13.1	13.9	14.2	15.2	16.7	17.0
9.0	12.7	13.0	13.4	13.8	14.8	15.0	16.1	17.6	18.0
9.5	13.4	13.8	14.2	14.6	15.6	15.8	17.0	18.6	19.0
10.0	14.1	14.5	14.9	15.4	16.4	16.7	17.9	19.6	20.0
10.5	14.8	15.2	15.7	16.2	17.2	17.5	18.8	20.6	21.0
11.0	15.5	15.9	16.4	16.9	18.0	18.3	19.6	21.6	22.0
11.5	16.2	16.7	17.2	17.7	18.9	19.2	20.5	22.5	23.0
12.0	16.9	17.4	17.9	18.5	19.7	20.0	21.4	23.5	24.0
12.5	17.6	18.1	18.7	19.2	20.5	20.8	22.3	24.5	25.0
13.0	18.3	18.8	19.4	20.0	21.3	21.7	23.2	25.5	26.0
13.5	19.0	19.6	20.1	20.8	22.1	22.5	24.1	26.5	27.0
14.0	19.7	20.3	20.9	21.5	23.0	23.3	25.0	27.5	28.0
14.5	20.4	21.0	21.6	22.3	23.8	24.2	25.9	28.4	29.0
15.0	21.1	21.7	22.4	23.1	24.6	25.0	26.8	29.4	30.0
16.0	22.5	23.2	23.9	24.6	26.2	26.7	28.6	31.4	32.0

1982 Tax Year

owners and maintains a market for the trust units so that owners can sell their units if they wish.

(*For more on the mutual funds specializing in government or municipal securities, see the next chapter.*)

Probably the most important consideration in deciding whether or not to purchase a tax-exempt security or fund is your income and tax bracket. The higher your income the more the tax-exempt status will bring you in extra yield. The accompanying table shows what tax-exempt bonds can mean to individuals and couples in various income brackets.

Corporate Bonds

In addition to the billions of dollars invested in government and municipal bonds, there is also a huge and active market for corporate bonds. These debt instruments are issued by the nation's business organizations to finance a variety of activities, such as acquisition of new plants and equipment, research and development, introduction of new products and the like. Some bonds are more speculative than others, depending upon the financial strength of the issuing company.

In order for the potential investor to evaluate or rate the relative safety of corporate bonds, two organizations—Moody's Investor Services and Standard & Poor's—continually rate these issues, as well as government and municipal securities, and if there is any significant change in the issuing company's financial position, it's reflected in either an upgrading or downgrading of the bond.

Should you purchase corporate bonds, you must remember that the issuing company can, under certain conditions outlined at the time of sale, call in or buy back the bonds after a certain period of time. This call provision enables the company to repurchase the issue from holders, and it comes into play frequently if the interest paid on the bond is well above the going rate. In most instances, when a company "calls" an issue it will pay a modest premium to the bond holder.

Another method of assuring a high return is through discount bonds. This is the name given to those bonds issued at a time when interest rates were lower. Since issuance, bonds that are now labeled "discount bonds" have declined so in price that the effective interest rate they pay is close to the rates paid by new issues. In addition to getting a high current interest rate on your discount bonds, you also will get more for the bond when it nears maturity, and that profit will be taxed as a capital gain. At that time the company will repurchase the bond at face value.

Of course, there is a caveat that goes with the purchase of discount bonds: Make sure the bond is selling at a discount because the initial interest rate (the so-called "coupon rate") is low—not because the company is in financial difficulties.

Another type of bond is the convertible. Convertibles are bonds that can be exchanged or converted into a specific number of common shares of stock of the issuing company, under terms and conditions outlined in the offering statement. Corporations issue convertibles in order to obtain a lower interest rate and eventually eliminate their debt, when all the convertibles are turned into common shares.

The conversion price usually is above the price of the underlying common stock at the time the bond is offered. Should the price of the common stock increase, the price of the convertible bond would rise accordingly. On the other hand, should the price of the stock decline, it is unlikely that the convertible would decline as sharply, because of the floor or base provided by the interest that the bond pays.

Like all other investments, the convertible is only as good as the company that issues it. So it is important to check such things as the company's sales and earnings. Remember, convertibles are another way of building your nest egg by investing in a growth company. When you own a convertible you enjoy the full benefits of corporate growth while at the same time receiving more income than you might obtain by owning the common stock. A useful rule of thumb—if the common stock isn't worth buying, then chances are the convertible bond isn't either.

"Things" and Other Things

Many financial planners believe that investing to build a nest egg should be free of speculation. But not necessarily. If you are careful and watch your investments such speculative investments as options, commodities, precious metals or art objects are worth considering. But one warning: It is vital that you have a good working knowledge of the subject you are dealing with before you even give the more speculative or nonconventional investments a thought.

It is most important that you know the value of what you have or are about to purchase. Auction houses are one important and open source of price information about art objects and antiques. Many auction houses publish price lists. If you're trading in metals, there are the newspaper financial pages and special publications devoted to those markets.

Precious Metals—Indeed, probably the most popular speculative investment in recent years has been precious metals, most notably gold. In 1977 the price of an ounce of gold was around $350. In the following three years gold soared to more than $850 an ounce. Then the bubble burst and prices plunged to below $400 per ounce. Some people made thousands in gold, some lost. Unless you want to bet on the market's short-term fluctuations, the value of gold as an investment at the present time is an open question.

If you do decide to invest in gold there are coins, ingots, jewelry, gold-backed bonds, certificates, mutual funds, gold mining shares of U.S. and South African companies, futures and options. The degree of risk varies. Following a price drop, South African gold mining stocks have provided a higher yield than most U.S. issues. Gold coins, including those minted by the governments of South Africa, Canada, Mexico and Austria, are readily available from banks and dealers but the dealer extracts from you a commission that could be as high as 10 percent, if you're making a small purchase. If you take delivery, don't store the coins in your home; it will cost you, but rent a safe deposit box or find some other method of safekeeping.

Silver is another popular precious metal. Since the early 1970s, prices have soared, plunged, soared and plunged again. You can invest in stocks of silver mining companies, pure silver coins, silver bullion or silver futures contracts. The risks of buying silver, either in its pure form or through shares of mining companies or mutual funds, are very similar to those of gold.

Diamonds, the popular song tells us, are a girl's best friend—but not necessarily a good investment. They are part and parcel of weddings and other ceremonies, and attractive to wear. Their value is not decreased by use and unless the wearer is very careless, a diamond cannot be damaged. Many people own diamonds and they remain another popular hedge against inflation. But they provide no income during the holding period and the only possible way you can make money is by selling your diamonds to another party willing to pay more than you paid. And this might be difficult if you bought them at a retailer's price markup. So you're better off putting your money somewhere else.

Options and Commodities—For those seeking to build a nest egg, stock options and commodity futures should be regarded as investments of substantial risk, with the possibility of substantial rewards—or substantial losses. An option gives you the right to buy or sell a particular security at a fixed price within a fixed period of time. The price of the option is always less than the price of the security represented by that option. If the price of the underlying security fails to move in the direction you expect it to in the designated period of time, you stand to lose your investment in the option. If the stock does go as you anticipate, you have the opportunity to gain disproportionately. If you believe the price of the underlying security will advance, you would purchase a "call" option. If you think there will be a decline in price, you would buy a "put" option.

Options for many securities are traded on the Chicago Board of Options Exchange (CBOE) and the American Exchange, and also on the Philadelphia and Pacific regional exchanges. Information may be obtained from your registered representative or brokerage house, which can also handle your purchases.

Even more speculative than stock options are commodity futures. Probably the most important thing to know about commodity trading is that if you don't have a solid knowledge of the commodity market and a good broker you can lose a great deal of money. Commodities typically traded on the commodity exchanges are corn and other grains, pork bellies, cotton, frozen orange juice, coffee, sugar, cocoa and the like. If you buy a commodity, you don't have to take delivery of a carload of pork bellies, say. What you get is a piece of paper showing your ownership interest in a given amount of the commodity in a warehouse or to be shipped. There is usually a sizable minimum investment.

A newly invented instrument is the stock-index future, which allows an investor to speculate on the movement of the market as a whole, as measured by an index such as the New York Stock Exchange Index, or the Standard & Poor's Index of 500 Stocks. Your stockbroker can give you information.

Neither options nor commodities are for the faint of heart or for those who don't have ample time to follow their investments. Changes come quickly, and unless you can tend to your holdings, this type of investment could prove a costly venture. By all means, don't invest more than a small portion of your nest egg in these markets.

Art and Collectibles—Investments also may include paintings, prints, furniture and antiques, rare books, flatware, crystal, and the like. The freedom to sell when you need to is much more limited than for stocks and bonds. There is no open or wide market for these items. You can only sell a valuable painting or rare stamp if there is someone else willing to buy. While you may get excited about the newspaper reports of record auction prices for Old Masters, you should also check out the many items that went unsold because there were no bidders at the minimum offering price.

There is an even more speculative group of "collectibles" such as antique cars, baseball cards, dolls, toy soldiers, stamps, autographs, pipes, jewelry, World War I and World War II airplanes, fine wines, beer cans and bottles, Hummel figurines, model trains, matchbooks and posters.

While art and collectibles have attracted a great deal of interest, they are worth only what someone else is willing to pay for them. Also, collectibles earn no interest while they are being held. For the most part they are not easy to sell. Unless you are an expert on the subject, it is easy to overestimate—or underestimate—the worth of an object or collection.

There is nothing wrong with attempting to use collectibles as a portion of your nest-egg-building program. But always buy any collectible item for your own enjoyment or use first, and as an investment second. After all, the market even for fine art fluctuates. You must ask yourself whether you'd be happy if you found yourself "stuck" with an Andy Warhol.

Real Estate—Real estate, defined by Webster's as "property in buildings or land," is an important asset for most American families. Investment property can take many different forms. It includes raw land, homes and apartments and commercial property. Ten years ago real estate investment trusts (REITs) and real estate syndicates became popular. But there can be serious dangers in investing in REITs or syndicates. The fluctuations of the economy and volatile real estate markets, coupled with inflation, high interest rates, and the bankruptcies of many commercial investment syndicates, have buffeted REITs and a number have suffered catastrophic losses.

There's a similar danger should you purchase individual commercial properties. The time and the place are vital. A large number of properties in all sections of the country have been hurt or bypassed by shifts in business or population, dropping substantially in value as a result. A new highway, for example, can benefit some properties, and draw people and money away from others.

Residential property, whether it be a single-family house or an apartment complex, can be purchased for rental income. Such factors as location and the economic conditions in the area are crucial in buying this type of property.

Over the long term, an investor can buy real estate to provide income and capital appreciation, as well as for the tax advantages that it offers. On the other hand, there are a number of problems that should be noted. Although it's easier than with a collectible, it's not possible to sell a parcel of land, a rental property or a commercial building at a moment's notice. First, you must find someone who is willing and able to buy and then is willing to pay the price you are asking. That takes time. If you are pressed to sell, this could be a costly problem.

One other important warning! Real estate has to be managed. Absentee ownership often can prove fatal in a real estate operation. If you can't see your property at least once a year, don't own it.

There is another—much safer—form of real estate investment: the ownership of your own house, cooperative apartment or condominium, purchased primarily as a home for you and your family. Currently, the most common investment is the ownership of a house. But it wasn't always that way. Until 1947, people who owned their own home were in the minority. Since the end of World War II, that has changed and it is now estimated that more than 60 percent of the nation's families live in dwellings they own. The site no longer has to be in the suburbs. Over the last 20 years, brownstones, co-ops, condominiums and manufactured housing (or mobile homes as they were called until recently) have all become popular methods of owning your own home.

Whether or not a person should own his or her own home depends upon the individual. Some people prefer to live in metropolitan areas and are totally unsuited for ownership of a house. For that group—and it is a large and growing one—cooperatives and condominiums provide many of the same advantages. These dwelling units are within buildings or complexes that are managed and maintained by professionals or by the owner-tenants as a group. For tax purposes, the individual units are treated the same as separate houses, and they may be bought and sold in the same manner as houses. If you are considering buying your own home, don't forget that if you sell after you're 55 the first $125,000 of capital gain is tax-free. But to qualify you must have used your home as your primary residence for at least three of the last five years before you sell it.

Protecting Your Nest Eggs

Whatever investment you finally select—whether you put your eggs into one basket or a number of baskets—it's vital that you watch your holdings. Because economic and business conditions change, you cannot forget about your investments. If you do, there's a good chance you'll be wiped out.

Never, never decide you are "just too busy" to look after what you have. The money you have put away for your nest egg must be watched. Like a garden, it's not likely to thrive if left to itself. It needs the green thumb of the gardener. Know when to plant—and when to prune. If you do, your chances are good for a bountiful harvest.

CONSUMER TIPS

Keep Your Nest Egg From Breaking

- ☐ Take an active, not a passive, approach to protecting your nest egg. Your savings for the future are at the center of your financial well-being.
- ☐ Keep this basic idea in mind at all times: the elements of your nest egg, taken together, must enable your money to grow faster than the rate of inflation, after you've paid the taxes on the growth.
- ☐ Make a list of your personal financial goals.
- ☐ Look for nest-egg plans and sources of advice that offer the best combination of safety and highest rate of return.

☐ Consider many possibilities and be sure to weigh the trade-offs involved.

Here are some of your options:

CERTIFICATES OF DEPOSIT
☐ Offer relatively high rates of return.
☐ Are federally guaranteed.
☐ Require infrequent decisions and demand little attention.
☐ Are worthy of consideration as a key element of your nest egg.

SAVINGS ACCOUNTS / "NOW" ACCOUNTS
☐ Pay interest, but usually at a rate far lower than the inflation rate.
☐ Provide easily accessible emergency funds.
☐ Are good for short-term savings.
☐ Should not be used as a storage place for the bulk of your nest egg.

MONEY MARKET DEPOSIT ACCOUNTS
☐ Provide instant cash for emergency use.
☐ Require minimum investment of $2,500.
☐ Pay interest at significantly higher rates than those paid to regular savings accounts.
☐ Are insured by FDIC and thus are safer than money market funds.
☐ Allow interest rates to vary, so are worth watching.

COMMON STOCKS
☐ Demand that you have a clearly planned program of investment.
☐ Require that you take a very active role in building a nest egg, monitoring your stock's activity carefully and making frequent decisions based on your own needs, abilities, age and family status.
☐ Work most successfully when your strategy is to spread your money over a carefully balanced combination of growth stocks, income stocks and total return stocks.

PREFERRED STOCKS
☐ Pay a fixed return or dividend.
☐ Are priced in line with corporate bonds rather than with common stocks.

Do Your Homework

Keep track of what's happening with your portfolio, no matter what type you've selected.
☐ Keep up with your bookkeeping.
☐ Know when to sell as well as what to buy.
☐ Follow the market daily or at least weekly.
☐ Read quarterly or annual reports.
☐ Scan the financial section of your newspaper.
☐ Consider subscribing to letters provided by investment advisory services.
☐ Seek an investment counselor if you have a large portfolio and little time to give it proper attention. Paying the fee will be worthwhile.
☐ Call your broker for information or to ask questions.

HELP IS HERE

Remember, there are two routes available to you if you are uncomfortable making stock selections on your own and do not want to make use of a mutual fund.

- Investment Clubs
    - ☐ Allow you to take an active role in deciding on investments.
    - ☐ Provide advice on important decisions.
- Brokers / Money Managers
    - ☐ Require fairly inactive role on your part as an investor.
    - ☐ Require your complete confidence in your chosen agent.
    - ☐ Need annual reappraisals.

ARE BONDS THE WAY TO GO?

Keep bonds in mind as still another alternative. They represent a corporate or government IOU:

- U.S. Government Bonds
    - ☐ Are the safest type of bond.
    - ☐ Include Treasury bills (taxed only by federal government) and bonds as well as obligations issued by federal agencies (taxed at state, local and federal levels).
    - ☐ Can be purchased in denominations of $5,000 or $10,000.
- Municipal Bonds

The most important consideration in deciding whether or not to purchase a tax-exempt security or fund is your income and tax bracket. The higher your income, the more tax-exempt status will bring you in extra yield.

- ☐ Short-term issues or notes usually require a minimum investment of $25,000.
- ☐ Long-term municipals require a minimum investment of $5,000.
- ☐ Municipals are tax-exempt.
- ☐ Can also be purchased through investment trusts or open-end mutual funds specializing in government or municipal securities.

- Corporate Bonds

Here's how to assure a high return when investing in a corporate bond.

- ☐ Evaluate the bond carefully before you invest. Consult its rating in **Moody's** and **Standard and Poor's**.
- ☐ Consider buying discount bonds. But be sure the bond is selling at a discount for the right reason (low interest rate) and not for the wrong reason (because the company is having financial difficulties).
- ☐ Take a look at convertible bonds, the kind which will be exchanged later for shares of common stock. This is another way of investing in a growth company. But keep in mind that if the common stock isn't worth buying, then neither is the convertible bond.

Speculative Investments

Speculation can be part of your investing to build a nest egg. But you must acquire a working knowledge of speculative investments before you even consider placing your money in them. Here are possibilities and precautions to keep in mind when you consider other "things."

PRECIOUS METALS
- ☐ Be prepared to pay a 10% commission to your dealer.
- ☐ Rent a safe deposit box or find some other method of safekeeping; don't store gold or silver in your home.

DIAMONDS
- ☐ Provide no income while you're holding them.
- ☐ Should not be a major part of your nest egg.

OPTIONS AND COMMODITIES
- ☐ Are highly speculative as investments.
- ☐ Require time and knowledge on your part as an investor.
- ☐ Should play only a small role, if any, in your nest egg building.

COLLECTIBLES
- ☐ Are speculative.
- ☐ Earn no interest.
- ☐ Are generally not easy to sell.
- ☐ Should be purchased for your own enjoyment and not primarily as an investment.

REAL ESTATE
- ☐ Can be highly speculative.
- ☐ Must be managed carefully.
- ☐ Is not always sold easily.
- ☐ Can play an important part in nest egg building if real estate owned is your own home.

Always make time to watch your nest egg and take steps to help it grow. This is crucial no matter which types of investments you choose.

CHAPTER 21

LETTING THE EXPERTS DO IT

If you don't have the time or don't want to go through the bother of supervising your investments and making yourself an expert on the market, you should invest instead in a mutual fund. There are plenty of these funds, some holding special types of common stocks, others money market instruments, still others such things as municipal bonds.

Mutual funds provide an easy method to build savings if you have only a limited amount of money to invest. And the beauty of it is you turn the whole task of investing over to groups of real experts.

There's nothing mysterious about these experts. They are professional investment managers whose judgment commonly is backed by security analysts and other specialists, and their decisions are subject to review by the fund's board of directors or its investment committee. You can find out who the leading managers are through publications like Forbes, Barron's and Financial World. For example, Barron's not long ago published profiles of a half-dozen leading fund managers. The star performer for the year was described as a tireless legman, who made it his practice to visit personally each company whose stock was held by the fund, and to interview at length the company's top management. He had a far more intimate knowledge of the companies behind the stocks and bonds than any individual investor could accumulate.

When the Fosdicks were trying to select a mutual fund to build their savings after their son was born, one of the terms they came across in virtually all the literature they received was "professional management." They learned that the people who run mutual funds are usually managers or management firms hired by the fund's directors. They discovered that the managers and the analysts that work for them read a variety of trade publications, attend meetings with industry and company officials, visit plants and facilities of a company whose stock they are interested in purchasing, and often have access to senior management people should a question arise.

Often the analysts and fund managers must make their recommendations to buy (or sell) a stock to a committee of senior fund management people. The Fosdicks learned that meetings of this sort are held frequently and the questions asked often are searching and detailed. In the course of doing their research on a stock or an industry, analysts and fund managers travel thousands of miles all over the country—and even abroad—and visit hundreds of companies in search of the latest and most useful information.

One of the things that pleased the Fosdicks about the funds they were considering was that once the stock is purchased, it is not forgotten. The fund manager keeps close tabs on the company (and the industry) and if there are significant changes, quick action is possible. This includes both purchasing additional shares or, if the news is bad, liquidating or selling some holdings.

Of course, the Fosdicks also found out that even with the most astute and detailed effort and the most careful watching of industry and company trends, there is no way to guarantee that the stocks in the fund's portfolio all will do well. But, whether for the Fosdicks or you, the benefits of professional management are so notable the purchase of a mutual fund can be financially rewarding over a period of years. And, by leaving the bulk of the homework to professionals, you save a lot of worry.

Mutual fund managers differ in their investment philosophies as well as in their methods. Yet you can choose from several fairly standard approaches that have evolved over the years.

The mutual fund idea began in London in 1868, when the "Foreign and Colonial Government Trust" was formed. The purpose of the fund was to provide ". . . the investor of moderate means the same advantage as the large capitalists, in diminishing the risk of investing in Foreign and Colonial Government Stocks, by spreading the investment over a number of different stocks." Today that basic concept has not changed significantly.

Initially, all were organized to invest shareholder funds in stocks or bonds paying interest or dividends. But as more funds came along and strove to attract their share of investor money, new approaches and investment strategies came about.

Funds have been tailored to produce such things as growth, income, growth and income combined, aggressive growth, and tax-free income. The funds' investments now include convertible securities, preferred stocks, municipal bonds, options, tax-exempt securities, government bonds, money market instruments and even precious metals, commodities and foreign securities, as well as the security they started with—common shares.

For instance, here's a look at all the different funds offered by one management organization. This firm, like many others, has adopted a "shopping cart" or "supermarket" approach to the funds under its wing, offering such investment vehicles as:

Stock Fund—invests in equities for a combination of growth and income.
Bond Fund—invests in a diversified group of bonds for high income.
Growth Fund—invests for long-term stock appreciation.

Total Return Fund—invests for high total return without restrictions as to types of securities.

Preferred Fund—invests in a diversified group of preferred stocks.

Income Fund—invests for income and conservation of principal.

Balanced Fund—invests for income, growth and conservation of principal.

Cash Reserves—a money market fund.

Government Reserves—a money market fund investing in U.S. government securities.

Tax-Free Money Fund—a money market fund providing a tax-free yield.

Tax-Exempt Bond Fund—invests in municipal and other tax-exempt bonds.

Capital Gain Fund—invests for maximum capital appreciation and takes higher risks for higher rewards.

So the investor in any fund in this family of funds has a wide choice of optional vehicles—all managed by the same firm. In addition, you can receive such services as automatic reinvestment of dividends, monthly withdrawal plans, Individual Retirement Accounts and Keogh plans.

Another management firm offers additional funds specializing in established growth companies, natural resources, small growth companies and stocks and bonds of non-U.S. companies. And still another firm has included a fund investing in shares of drug and health-related companies and a fund dealing solely in stock options.

Of course, these are only examples of what is available from many mutual fund organizations. What all funds provide is:

- Professional management;
- Diversification of risk;
- Free access to information for all shareholders;
- Freedom from day-to-day worry;
- Freedom from record-keeping (quarterly statements are available from most mutual funds);
- Automatic dividend reinvestment plans, or the option to receive dividends in cash;

- Instant diversification;
- Unlimited exchange privileges from one fund to another within a management group;
- Ready redemption of shares upon request.

When making a decision regarding what type of fund to buy, you are faced with making a basic choice between a fund with a "load," or sales charge, and a "no-load" fund, one without a sales charge. The sales charge on some ranges as high as 8.5 percent of each investment. Thus if you send $100 to the mutual fund organization, only $91.50 will be invested. The remainder goes to the management firm and the selling agent.

No-load funds are sold directly from the fund sponsor to you, the buyer, without a sales charge. Here, if you send the fund $100, the entire amount is invested in securities. The cost of operating the fund is later charged against the value of the fund's assets—which is also the case with load funds.

Whether or not the fund charges a load does not affect its performance. Both types provide professional management. Among both the load and no-load funds during the past 40 years there have been winners, and losers.

With a sales charge, or load fund, you do get the counsel of a securities representative. If he or she is a professional, that could be very helpful to many investors who otherwise would be uncomfortable investing by mail, without a personal contact. To buy a no-load fund you must take the initial step of contacting the management firm rather than visit your stockbroker.

Open-End vs. Closed-End Mutual Funds

If you decide you want to invest in a mutual fund, you will find that shares of virtually all funds are readily available. Yet there is a significant difference in how they're obtained, depending on whether the fund you want is "open-end" or "closed-end."

Open-end means that the fund's management stands ready to sell you shares immediately upon receiving your order. If, for example, there is a sudden influx of new orders, an open-end fund will create as many shares as necessary to satisfy public demand. The shares always are sold at net asset value—that is, total value of all investments and cash on hand, divided by the number of shares issued, giving value per share. Plus a sales charge, if any. Conversely, if investors decide to redeem their shares, the open-end fund will accept the orders and repurchase all the shares the public has offered. This feature allows you to purchase or sell your shares without regard for what anyone else does. The new sales do not dilute the net asset value, as the money goes immediately to expand assets, either as cash on hand or as new investments. Nor do redemptions affect per-share value.

While open-end funds claim they stand ready to sell new shares or redeem existing shares at all times, there have been situations where they have had to back away from that promise. For instance, an open-end fund can become a favorite of many buyers and receive more money than it can properly handle. This sudden influx of new money could cause the fund to close itself to new shareowners for a period of time, or else become a closed-end fund. Should the latter event take place, a new shareowner could only purchase shares from a seller and the transaction would have to be handled by a broker. In recent years, with the advent of money market funds, particularly those catering to institutional accounts, the number of funds that have had to close their doors to new accounts has been very small.

On the other hand, situations could arise where a large number of shareowners wish to sell at approximately the same time. Such a situation could cause a run on the fund and result in the sale of some portfolio holdings should the fund be unable to meet redemption requests out of its available cash reserves. The need for the fund to liquidate stock holdings could cause market prices to fall. If the fund has large positions in stocks purchased at higher prices than those paid by the current market, such a run could result in a drop in value of the fund's shares.

Even with mutual funds, one has to remember that the stock market is a two-way street, with both ups and downs. As J.P. Morgan, when asked for his prediction on the stock market, once said: "It will fluctuate."

Closed-end funds have a fixed number of shares outstanding, or available to the public. If you want to buy shares of a closed-end fund, the majority of which are listed on the New York Stock Exchange, you can place your order with a stock broker and wait until someone else wants to sell. In most instances there is a ready market on both the buy and sell side and orders generally are filled quickly.

There is one thing to remember. When buying and selling shares of a closed-end fund or investment company, you must pay a commission to your stock broker. Another factor in buying closed-end funds: There may be times when the shares sell substantially below asset value. The fund's shares are traded on the market just like any corporate stock.

In contrast, the open-end funds are priced according to the value of assets owned. So, the price of the closed-end fund may depend on the public's interest in the fund rather than the underlying value of its assets or share holdings. At other periods of time, the shares of a closed-end fund can be priced substantially above the value of all its holdings. So at times when prices are at a premium you need to take special care when considering a closed-end fund. Closed-end funds in recent years generally have been priced at a discount. However, this does not mean an automatic gain for you, as the fund may still be at a discount when you decide to sell.

Money Market Funds

The biggest success story of the mutual fund industry between the mid-1970s and the early 1980s was the growth of money market funds At the end of 1974, there were 15 such funds available, with assets of slightly more than $1.7 billion. By late 1982, when declining interest rates were lowering their appeal somewhat, there were more than 200 money market funds with combined assets totaling well over $200 billion.

Portfolios of these funds consist primarily of short-term, low-risk securities, such as U.S. Treasury bills, high-quality commercial paper (corporate IOUs), and bank certificates of deposit.

The popularity of money market funds has been based on their relatively high rate of return, safety of principal and ease of withdrawal. The vast majority of money market funds do not charge a sales fee. Another option for investors in money market funds is the privilege of writing checks against their balance. Usually the withdrawal must meet a specified minimum, typically $500.

The primary disadvantage of money market funds versus Treasury securities or bank certificates of deposit is that they are not federally guaranteed.

Although they are subject to taxes, money market funds offer a number of attractive features that you should consider, such as withdrawal of funds without penalty, small initial investment (as low as $1,000), no sales charge, no time requirement to earn maximum yield, high current yield, ease of addition and withdrawal of funds without penalty, relative safety and low expenses.

Selecting a Fund

With the introduction of money market funds, the task of selecting the right fund has become even more difficult. But various fund organizations have made the choice a great deal easier by establishing exchange privileges. Under this program, organizations with more than one fund under their wing allow any shareholder to transfer his investment from one fund to another within the same family to accommodate changes in needs or objectives or to take advantage of market changes. Usually, there is a nominal administrative fee for switching from one fund to another.

Certainly the first thing you should do is choose your own investment objective. Decide whether you want growth, income, or a combination of both. You must choose among portfolios consisting largely of common stocks, convertible bonds, senior securities or tax-exempt bonds. You should be aware of the record of each fund that you are considering in all types of markets—bull, bear and boar (when a market just lies inert). Also, you need to know about a fund's specific portfolio, its management philosophy, its operating costs, its sales charges, and the services it will offer.

Most of the information you seek is available in the prospectus and the annual report of each mutual fund. In fact, you should not buy any fund before you have studied the latest prospectus. The Securities and Exchange Commission requires that it be provided to you. At your request, every mutual fund, regardless of size or objective, will mail you the material you need.

Once you have the prospectus and the annual reports, it is worth your time to read them carefully. Look at the portfolio of holdings, understand the investor services provided and know whether or not there is a "family" of funds available that will allow you to switch later to another fund with another objective.

One method of judging any mutual fund's performance is by looking at the historical record. Usually the annual report presents this in the form of a graphic illustration of an assumed investment of $10,000 or one share, with income and capital gains distributions accepted in additional shares. The chart may start with the individual fund's inception and go through the latest reporting period, or it may cover other periods.

Another method for portraying a fund's performance is by showing how the fund acted over various 10-year periods. In most such cases, you will find that the period covered is one of rising common stock prices; if the fund invested primarily in such securities, the graph will show an uptrend. All such examples carry the warning: "The results shown should not be considered as a representation of the dividend income or capital gain or loss from an investment made in the fund today." In other words, don't expect too much.

In order to project a fund's future performance, it's a good idea to look for a past period when conditions paralleled those of the present. For instance, past periods of rising prices, or "bull markets," should not be compared with recent "bearish" or downward trends.

After you have selected your fund and made your purchase, you should continue to keep track of the net asset value. Most of the nation's leading newspapers provide these prices at least once a week. If not, nearly all mutual funds have shareholder service departments that will provide you with the necessary information.

The Magic of Dollar Cost Averaging

If you're like most American wage earners, you accumulate money on a gradual basis over an extended period of time. You only read about the "big winners"—those people who are favored by a lottery or an inheritance.

Since most people will not fall into the "big winner" category, the amount of money available at the end of a month or a quarter is limited—and that money has to go a long way. Probably the best method for accumulating your nest egg if you can only put away a small amount regularly is by "dollar cost averaging." And if you've chosen to invest in an open-end mutual fund, it's an easy and virtually painless method that will enable you to build and maintain a meaningful investment program for whatever goal you hope to achieve.

But bear in mind that unless you are accumulating shares of a no-load mutual fund, a commission or sales charge will be added to each purchase you make.

Simply put, dollar cost averaging is the purchasing in fixed dollar amounts ($50, $100, $250 for example) of shares at regular intervals (monthly, bimonthly, quarterly), regardless of the prices of the stock or fund at the time of purchase. When the procedure is followed without interruption, you will obtain more shares when the price is low and fewer shares when the price is high. Thus, given normal fluctuations in price over an extended period, the average cost to you of all your shares will be less than the average price during the period. If you follow a program of dollar cost averaging faithfully, you can accumulate shares at a reasonable cost—regardless of the price at any one time.

That's what the Fosdicks did. They decided to invest a fixed amount of money—$200—in a fund on a regular quarterly basis. Here's the record of their purchases for five consecutive quarters:

Amount Invested	Share Price	Shares Purchased
$200	$10	20
200	8	25
200	5	40
200	4	50
200	8	25

Shares acquired through $1,000 investment   160
Average price per share ($35 divided by 5) = $7.00
Average cost per share ($1,000 divided by 160) = $6.25

Thus, even if the shares had been redeemed at only the average price of $7, the Fosdicks would come out ahead by 75 cents per share. The longer such a program goes on, the more you will benefit from this quirk of arithmetic.

While the range of prices is overstated in the example of the Fosdicks' dollar cost averaging program, nevertheless the principle holds true: The same dollar amount accumulates more shares when prices are low than when prices are high.

There is a special lesson from all this. Don't be guided by emotions when making purchases. Regardless of the price of the fund at the time payment was due, the money was sent. The Fosdicks sometimes found that hard to stomach, especially when the price of the fund was well below the original purchase price. They feared they were putting good money after bad, but over the long run they benefited from the price dips, and the policy proved worthwhile.

Nearly every mutual fund has an accumulation program, by which investments on a regular basis are accepted. When combined with a program of automatic reinvestment of all income and capital gains distributions, a modest regular investment can add up to substantial growth.

Of course, as is the case with all investments, the fund must be sound and well managed. As all mutual fund organizations state in their literature when discussing dollar cost averaging: "A program of this type does not assure a profit or protect against depreciation in declining markets." But it can improve your chance of gain—mathematically.

Reinvesting Your Gains

Whether or not you choose the dollar cost averaging approach, you should also be aware of the value of reinvesting both dividend income and capital gains distributions. Mutual funds obtain dividends and interest from their investments. After deducting operating expenses, the remainder is paid out to shareholders. Payments can be made annually, semiannually or quarterly, depending largely on the size of the fund and the amount of the distribution. Most mutual funds make income distributions on a regular basis.

A mutual fund will realize capital gains when the value of securities sold at a profit exceeds those sold at a loss. Companies may either retain the profits and pay taxes on behalf of the shareholder or pay out the gains and have you, the shareholder, pay the taxes. In most cases, funds will follow the latter policy. Capital gains distributions normally are made on an annual basis.

Although the requirements of each investor are different, if you reinvest both income dividends and capital gains distributions you can make the amount of your holdings grow far more substantially than if the distributions had been taken in cash. Of course, situations can arise where you may find it necessary to take the distributions. Normally, though, your nest egg will show far more rapid growth if all dividends are reinvested, because of the magic of compounding—another mathematical quirk that works for you. Compounding means that you are earning dividends or capital gains on a principal amount that is larger after each reinvestment—making your investment grow at a geometric rate.

Caring for Your Mutual Fund

The fact that you have purchased a mutual fund and left to experts the decisions on what to buy or sell is no reason for you to forget about your savings. Your money is too important and too vital to treat in a casual manner.

Probably the most important thing you can do to monitor your fund's performance is to read the annual report and the semiannual or quarterly reports that the management provides. This material generally discusses economic conditions, reviews the fund's changing portfolio, and compares the fund's performance to various stock market indicators (including the Dow Jones industrial average and the various Standard & Poor's indexes). When reading the fund's reports, you should note whether the fund has outperformed the various averages and whether the portfolio consists of the kind of securities you are interested in owning.

You also should check your fund's performance through your newspaper (most Sunday financial sections show the weekly price movement of all mutual funds). Also, The Wall Street

Journal, on a daily basis, and Barron's, on a weekly basis, provide useful information to help you track the industry. Annually, Forbes magazine reviews all mutual funds, and its performance ratings are a useful guide. Money magazine also provides useful information on the industry.

In addition, there are various information services, available on a subscription basis or in brokerage houses or public libraries, that can be of help.

In no case should you leave your investment fund alone. That can be a serious and costly mistake, as scores of purchasers have found out, especially those who went into "go-go" funds in the 1960s only to discover later that the "go-go" had gone.

If your fund's performance does not match the averages or that of other similar organizations, do not hesitate to review your investment in it, and to consider a change. Whether you own a load or a no-load fund, all shares are redeemed at net asset value. (That's the price that's given in the newspaper financial page listings.) There is no penalty or redemption fee.

You worked hard to accumulate your nest egg, and it is worth your attention to see that it doesn't dwindle.

CONSUMER TIPS

Mutual Fund Facts

- ☐ They can help you to build a nest egg, even if you have only a limited amount of money to invest on a gradual or periodic basis.
- ☐ Mutual funds provide professional management for your investments and are organized to invest your funds prudently.
- ☐ They are tailored to meet specific financial goals.
  Examples:
  - ☐ Growth
  - ☐ Income
  - ☐ Growth and income
  - ☐ Aggressive growth
  - ☐ Tax-free income
- ☐ Mutual Funds differ according to the approaches and investment strategies of the fund management.
- ☐ You are offered a wide choice of investment vehicles, all of which are managed by the same firm.
- ☐ These are some additional services:
  - ☐ Automatic reinvestment of dividends
  - ☐ Monthly withdrawal plans

- [ ] Individual Retirement Accounts
- [ ] Keogh Plans

LOAD OR NO-LOAD?

A basic choice to be made when considering mutual funds is deciding between a fund with a "load" and a "no-load" fund.

"Load" Funds

- [ ] Provide you the counsel of a securities representative.
- [ ] Levy a sales charge of as much as 8.5% of each investment.

"No-load" funds

- [ ] Are sold directly to you from the fund sponsor.
- [ ] Charge the cost of operation against the fund's profits.

Either way, you pay, and the past performance records of mutual funds appear not to have been affected by the "load" or "no-load" factor.

DECISIONS, DECISIONS

A second decision you'll need to make concerns how your shares will be obtained.

Open-end funds

- [ ] Allow you to buy or sell shares without regard for what anyone else does.
- [ ] Are always sold and repurchased at net asset value (total value of all investments and cash on hand divided by number of shares issued = value per share), plus a sales charge, if any.

Closed-end funds

- [ ] Have a fixed number of shares outstanding, or available for sale.
- [ ] Require the payment of a commission to your stock broker.
- [ ] Can sell, at times, substantially below or above asset value.
- [ ] Are traded on the market like any corporate stock.
- [ ] Must be considered cautiously when prices are high.

MONEY MARKET FUNDS

These were the most successful type of mutual fund during the past decade. Their most attractive features include:

- [ ] High rate of return
- [ ] Ease of withdrawal
- [ ] Small initial investment
- [ ] No sales charge
- [ ] No time requirement to earn maximum yield
- [ ] Relatively safe
- [ ] Low expenses

The primary disadvantage of money market funds is that they are **not** federally insured.

HERE'S HOW TO CHOOSE A FUND

- [ ] Select an investment objective.
- [ ] Read the latest annual or quarterly reports and prospectuses for the funds which interest you.
- [ ] Look at the funds' historical records of performance.
- [ ] Decide which fund best suits your personal needs. (Keep in mind that even with mutual funds, the stock market is a two-way street and thus has both ups and downs.)

Letting the Experts Do It

☐ Make a choice.
☐ Try investing a fixed amount of money in a fund on a regular quarterly basis—find out what "dollar cost averaging" can do for you.
☐ Reinvest both income distributions and capital gains dividends when you can afford it so your nest egg will grow as rapidly as possible.

HOW TO CARE FOR YOUR FUND

☐ Monitor its performance carefully
☐ Read annual and quarterly reports to see how your fund's earnings record compares to others.
☐ Check daily or weekly in newspapers and financial magazines to see how your fund is doing.
☐ Consider subscribing to various information services or read newsletters in your public library.

As with all investments, take an active role in caring for your mutual fund. It's a necessary part of protecting what's yours.

CHAPTER 22

RETIRING WITHOUT PENSIONING OFF WHAT'S YOURS

The main purpose of a savings plan is to amass enough money so that you can live comfortably when you retire and not have to dig into the capital you've built up or sell your home or your personal possessions. At the same time, you want to be sure the money you are building up in your investments and retirement plan is sheltered as far as possible from the clutches of the tax collector. As a result you not only need to think now about how to save but also about how much you will need when you retire. And you need to think ahead about ways to supplement your income once you end your full-time career. If you don't, 10 or 20 years from now you could find your possessions being whittled away for the simple necessity of living.

One expert suggests that people view retirement as a "graduation" either to second careers or the more active pursuit of hobbies and other interests they have developed during working years. Since most people view their retirement income prospects as inadequate, and fear the ravages of inflation on the income they will have, he advises people to think ahead or work now to develop sources of extra income from activities such as consulting or part-time work.

The range of retirement "jobs" is wide, and you have the option of doing something you enjoy. Retirement may be a perfect time for doing something you always wanted to do: Start a small business; run a small farm or a country inn; operate a resort; sell your paintings or photographs; do some writing or teaching. To have a real choice, you will need enough financial backup to relieve you from the pressures of necessity. And that's where your retirement financial planning comes in.

Although your expenses will be somewhat less after you retire, you will need about 80 percent of your pre-retirement income to live in the same style you're accustomed to. But, there are some factors that could affect that estimate.

Financial planning starts with your assets. Your home may be chief among them. Will your mortgage be paid off by the time you retire so that you will be able to live free of those burdensome payments? If so, that change will be as good as so much extra income. If your mortgage will still have some time to run, or if your house will represent a drain because of its size and the cost of taxes and heating bills, you may wish to use your built-up equity in your home to switch to a smaller house in an area where the climate is warmer and the taxes lower. If your children are grown and no longer living at home, you may not need as large a house, nor will you have the cost of their support—a big item for most families.

Once you have determined where you will want to live and about what your living costs will be, you can start to plan your finances intelligently. Even before you make these decisions, there is a lot of planning you can do.

To help you start thinking seriously about retirement, the following table shows what the life expectancy will be for men and women at various age levels.

Life expectancy at age 65, according to the table, is 16.7 years, for men and women combined. However, this is only an average. It would be a big mistake for you—or anyone—to base your retirement planning on making your funds last just so

Years of Life Expectancy in the United States

	White			All Other			Total		
Age	Male	Female	Total	Male	Female	Total	Male	Female	Total
0	70.6	78.3	74.4	65.5	74.5	69.9	69.9	77.8	73.8
1	70.5	78.1	74.3	66.0	74.9	70.4	69.9	77.6	73.8
5	66.6	74.2	70.4	62.2	71.1	66.6	66.1	73.8	69.9
10	61.8	69.3	65.5	57.4	66.2	61.8	61.2	68.9	65.0
15	56.9	64.4	60.6	52.5	61.3	56.9	56.4	64.0	60.2
20	52.3	59.6	55.9	47.8	56.4	52.1	51.8	59.1	55.5
25	47.8	54.7	51.3	43.5	51.7	47.6	47.3	54.3	52.8
30	43.2	49.9	46.6	39.3	47.0	43.1	42.8	49.5	40.6
35	38.6	45.0	41.8	35.1	42.3	38.7	38.2	44.7	41.0
40	34.0	40.3	37.2	31.0	37.7	34.4	33.6	49.9	36.8
45	29.5	35.6	32.6	27.1	33.3	30.2	29.2	35.3	32.3
50	25.2	31.1	28.2	23.3	29.2	26.2	25.0	30.8	28.0
55	21.2	26.8	24.1	20.0	25.4	22.7	21.1	26.6	23.9
60	17.5	22.6	20.2	17.1	21.7	19.6	17.5	22.5	20.1
65	14.3	18.7	16.7	14.6	18.5	16.6	14.3	18.7	16.7
70	11.4	15.1	13.4	12.0	15.2	13.7	11.4	15.1	13.5
75	8.9	11.8	10.6	10.1	12.8	11.6	9.0	11.8	10.6
80	7.0	9.1	8.3	9.7	11.8	11.0	7.2	9.3	8.5
85 and Over	5.6	7.1	6.6	8.9	10.7	10.0	5.9	7.3	6.8

Note: Data are provisional
Source: National Center for Health Statistics, U.S. Department of Health and Human Services

many years. You might live a lot longer, and it's even less fun being broke at 85 than at a younger age. So make plans that will assure you a continuing income—no matter how long you live.

Living costs can vary considerably depending on location. The following table shows what it cost a retired couple—a man over 65 and his wife—to live adequately in 25 different major U.S. cities during 1981.

The Cost of Retirement

What it costs a retired couple to live in:

Anchorage $12,900	Los Angeles area $10,238
Honolulu $12,157	Minneapolis-St. Paul $10,121
Boston $11,925	St. Louis area $10,108
New York area $11,623	Chicago area $10,070
Seattle area $11,343	Baltimore $10,051
Washington, D.C., area $11,000	Cincinnati area $10,038
San Francisco area $10,921	Denver $10,028
Buffalo $10,744	Houston .... $ 9,996
Milwaukee $10,673	Kansas City area .... $ 9,978
Philadelphia area $10,646	San Diego .... $ 9,827
Pittsburgh $10,503	Dallas .... $ 9,768
Cleveland $10,500	Atlanta .... $ 9,516
Detroit $10,395	

Source: Labor Department

Social Security

The bedrock of the retirement program for the vast majority of the American work force is Social Security. Established in 1935, the system currently covers nearly all employees of private business, most state and local government employees, household and farm workers, and members of the clergy, and also people who are self-employed. More than 36 million people now receive Social Security benefits.

Despite recent financial problems in the system, there is little doubt the government will go on supplying retirement benefits on into the future.

In order to receive a Social Security check each month, you must have credit for a certain amount of work covered. The following table shows in years how much credit you'll need in order to receive Social Security retirement benefits.

Work Credit for Retirement Benefits

If you reach 62 in	Years you need
1979	7
1980	7 1/4
1981	7 1/2
1982	7 3/4
1983	8
1987	9
1991 or later	10

Exactly how much Social Security you will receive each month will depend on your age at retirement as well as your past earnings. You can start collecting Social Security at age 62, but at a reduced rate. For an individual retiring in 1982 at 65, the top monthly payment was $729. A spouse is entitled to one-half the amount of his or her partner's monthly check. Thus, the top monthly benefit for a couple retiring in 1982 at 65 would be $1,093.50. At 62, these amounts are reduced 20 percent for workers, 25 percent for the spouse.

After 65, you currently are permitted to earn $6,600 a year from part-time work without reduction of benefits. As of 1983, earnings over $6,600 are deducted from your check at the rate of $1 for each $2 of earnings. This limit is scheduled to increase in future years, and may also be changed in new legislation.

To apply for Social Security, call or visit the nearest regional office of the Social Security Administration. Once you've been interviewed, either in person or by phone, the process can be finished by mail. Also, don't forget that your employer's personnel office may be able to help you.

The Social Security Administration has put forth various rules and regulations on such subjects as working past 65 or being able to receive Social Security if you should retire outside the United States.

This information is contained in Social Security booklets. So is information on hospital and medical insurance under the Medicare program—another important pillar of your retirement planning that is provided through the Social Security system. You can get these booklets and leaflets at any Social Security Administration office.

While most everyone can look forward to receiving Social Security benefits, a company pension upon retirement, or both, the amount may not be enough for your needs. Social Security was intended as an income supplement, not as a sole source of support. Even if you earn the maximum salary taxed for Social Security, and draw the maximum benefit, you could receive less than half your working pay. And you are unusually fortunate if you have a company pension that completely fills the gap between Social Security and what you need to maintain your lifestyle. So you'll probably want to supplement these programs with one of your own.

Corporate Pension Plans

Not all companies have retirement plans. Even more important, not all corporate pension plans are alike. While the Employee Retirement Income Security Act (ERISA) of 1974 includes a host of safeguards and provisions that should protect you, it also makes clear there is absolutely no uniformity among plans.

Thus, you should learn all the facts about your company's pension plan. Information normally will be readily available, covering such vital topics as type of plan, the conditions under which you become eligible for a pension, and the kinds of benefits available if you retire early, or at 65, or even later. The company is allowed to set its own rules, provided they are permitted under ERISA.

Two types of pension plans are available:

- A defined-benefit plan that promises a set benefit to you upon retirement.
- A defined-contribution plan, under which the company promises to invest a certain amount of money on your behalf, usually a percentage of salary, or set aside a percentage of profits (if any).

Some companies have both types of plans. Once you know what type of pension plan your company has, you need to know what the requirements are to become "vested," or eligible to receive a pension. Once again, every company has its own vesting arrangement. Under the provisions of the pension reform legislation, there are three options:

1. Fully vested after 10 years of employment;
2. 25 percent vested after five years of employment—reaching 100 percent vested after 15 years; and
3. 50 percent vested after your age and years of employment total 45, with vesting increased 10 percent annually after that until 100 percent has been reached.

Retirement age may vary, depending on the details of the plan. Some company plans permit early retirement with a partial pension, and some provide a full pension before age 65. The amount you finally do receive will depend on your years of employment and the salary you earned in the last few years of employment, and also how much Social Security you'll receive. Employers commonly deduct half your Social Security benefit from your company pension (since the company has put up half your Social Security contributions). You also should know whether your pension is taxable or not (the employer's contribution is taxable), whether there are survivor benefits and whether there are provisions for automatic increases in payments to offset the ravages of inflation.

So it is in your interest to read your company's plan publications and consult with the plan's administrator if you have questions. Although ERISA established the federal Pension Benefit Guaranty Corporation, which insures your vested benefits in a private pension plan, your company may change or terminate its plan. Also, if your company should merge or go out of business, pensions might be affected.

Thus, you should check out your company's pension plan and discuss it with the people in charge. You have a right to know where you stand.

Employee Savings Plans

A lot of workers have bolstered their retirement kitties by contributing to employee savings plans, which many companies have established. Such plans typically enable workers to place a certain part of their salary—generally around 5 percent—into common stocks, fixed income securities, money market funds or combinations of investment vehicles.

The employer also will make a contribution—usually from 25 percent to as much as 100 percent of the employee's contribution. If you're in a plan where the employer contributes 50 percent, for every $1 you contribute your employer will add 50 cents.

Employee savings plans are administered by a trustee and many people regard them as an ideal way to save. In many plans, you can make additional contributions but the employer will only match whatever maximum is stipulated in the plan. A savings plan provides an excellent opportunity to accumulate a hefty nest egg. There's one important plus. The money your savings earns, whether interest, dividends or capital gains, is not taxed until you get it. If you don't receive this money until after you retire, these earnings will usually be taxed at a lower rate than when you were working.

Rules of eligibility vary from company to company, but there usually are few restrictions. Should you need to withdraw the money it is available upon demand—no matter for what reason. The one drawback is that normally you must wait for a specified period before re-entering the plan. This restriction is written into virtually all employee savings plans in order to discourage withdrawals of small amounts of money.

The choice of where to put your money—should your savings plan have a number of options—is strictly up to you.

Among the questions to consider are the degree of risk you wish to take, what other savings plans you have and the purpose for which you are saving.

While employee savings plans are not available everywhere, they are becoming increasingly popular and provide an excellent method of accumulating a nest egg, especially if you find it difficult to save on your own.

Supplementary Retirement Plans

Congress amended the laws in recent years to make tax-deferred or tax-sheltered retirement plans more liberal and easier to obtain. Now they are available both to self-employed people and those who work for others. Probably the most sweeping revisions came with the Economic Recovery Tax Act (ERTA) of 1981. And even those changes were amended by the Tax Equity and Fiscal Responsibility Act of 1982. Congress eased eligibility requirements, and both Individual Retirement Accounts (IRAs) and Keogh Plans are now more attractive.

IRAs

The Economic Recovery Tax Act of 1981 made it possible for every wage earner to become eligible for an Individual Retirement Account (IRA). No matter who you are, thanks to ERTA you can now make a tax-deferred annual contribution of $2,000 or 100 percent of your annual earnings, whichever is less, into an IRA, and that amount is deductible from your current taxable income. If both you and your spouse work, you may each contribute $2,000 or 100 percent of your annual earnings, whichever is less, into an IRA. If you have a non-working spouse, the two of you together may contribute a total of $2,250 to your IRA and your spouse's IRA, but not more than $2,000 into either account.

While the increase in the annual contribution is welcome (it was $1,500 or 15 percent of your salary, whichever was less, prior to 1982), the most significant change was the eligibility requirement.

More important, you are now permitted to open an IRA even if you are already covered under a qualified company retirement plan that has been approved by the Internal Revenue Service. Until the end of 1981, you could not open your own IRA if you were covered by your employer's plan.

It's conservatively estimated that this change made more than 55 million workers suddenly eligible for IRAs. Add this total to the approximately 45 million previously eligible and it's little wonder that the nation's newspaper readers and television viewers have been bombarded with IRA advertisements.

The majority of the print advertisements show how your annual contribution will grow over the years, if current rates of interest continue. Of course, the ads do not guarantee that they will.

The choice of IRAs is staggering and almost every type of organization that handles money from the public now offers IRA plans. The list includes banks, mutual fund organizations, savings and loan associations, life insurance companies and brokerage houses. It should be noted that some of the organizations offering plans charge a fee for their services. These fees vary from as little as $5 for some mutual funds to $25 or more by some brokerage firms.

Fees for maintaining an IRA account do not count as part of your annual investment, nor does any commission charged by a brokerage firm for buying stocks or bonds (if you elect to open an IRA there). So you should be aware that such fees are an added expense that you may not be willing to absorb.

The Fosdicks became eligible to open their own IRA account in 1982 and found that the variety of organizations offering IRAs made the selection difficult. In fact, it was no easier than what they went through before they chose their mutual fund. Then they learned that they could turn their mutual fund account into an IRA, just by notifying the fund and filling out the necessary applications.

The Fosdicks, both of whom opened IRAs with the mutual fund, knew that annual contributions to their accounts would be tied up until they were $59^{1}/_{2}$ and could start receiving benefits without penalty.

You will have to give some thought to how much money you can invest annually (especially if you are just starting a career or a family). Once you open an IRA you are not obligated to make a contribution each taxable year, and you need not make the full $2,000 contribution if that becomes a burden.

There's a recent innovation you should know about. Many companies now allow employees to participate in IRAs through a payroll deduction program.

Whatever type of IRA you select, make sure you shop around. Find out as much as possible about the plan you choose. And make sure that at least once a year you receive a statement of the balance in your account—whether with a bank, an insurance company, a savings and loan association or mutual fund.

By all means, get into an IRA or some of the tax-sheltered savings plans. Otherwise the tax collector will grab so much of your savings there'll be little profit after inflation for you to use for your retirement.

Keogh Plans

The new laws also have liberalized Keogh Plans (pronounced Kee-oh). Keogh Plans are designed for self-employed people and their eligible employees (for example, a doctor and nurse, lawyer and secretary, or a business run by a small number of people).

The most important Keogh change was the increase in the annual maximum tax-deductible contribution. ERTA increased the annual contribution to $15,000 or 15 percent of yearly self-employed income, whichever is less. The 1982 act increased the maximum contribution to 20 percent of earned income or $30,000 for 1984. And beginning in 1986, the maximum dollar limits for Keogh Plans will be pegged to cost-of-living increases. Before ERTA, the maximum deduction had been $7,500.

Left unchanged by the 1982 legislation was the provision enabling a self-employed person to make contributions to both an IRA and Keogh so long as the combined contributions did not exceed the allowable limits. The act also repealed many of the special rules applicable to plans for the self-employed, including the requirement that the plan trustee must be a bank.

If you have a small amount of income from moonlighting, in addition to the salary you receive from your regular job, you can contribute to a Keogh Plan and shelter from taxes 100 percent of those earnings, or $750, whichever is lower. But there is an important catch. This provision applies only if your total adjusted gross income is less than $15,000.

When setting up a Keogh Plan, the same care should be taken as with an IRA. Remember that you can open more than one Keogh account, provided the annual contribution does not exceed the 15 percent or $15,000 limit through 1983. And remember, too, the overall attractiveness of IRAs and Keogh Plans as a method for reducing income tax while at the same time building interest or dividends year after year is hard to match anywhere.

Alternatives

Various programs designed by life insurance companies—endowments and annuities—also provide funds for retirement. Annuities are designed to give you or your spouse a specific amount of money each month for a set period or for the rest of your life. For the most part, annuities are not taxed. The exception is a plan that pays out more than your original contribution. As with all other retirement plans, there are many variations among annuities and you should become familiar with all the regulations governing your particular plan. Different types of annuities are discussed in Chapter 15.

Whether an IRA or an annuity, if you add one or more of these additional building blocks to your retirement plan to supplement your Social Security and pension payments, you should be able to look forward to a retirement where you can keep up your current life style and not have to part with any of the things that are yours.

CONSUMER TIPS

The Path to Retirement Security

Keep in mind the key elements that can move you toward a happy and secure retirement:
- ☐ Plan intelligently while you're earning.
- ☐ Be aware of your personal goals and your means of achieving them.
- ☐ Recognize and plan to minimize threats to your financial security, like inflation, and emergencies such as disability of breadwinner.
- ☐ Remember that sufficient financial backing can provide the kinds of choices you'll want to be able to make as a retiree. What could be a possibility for you?
    - ☐ Launching a second career
    - ☐ Pursuing your favorite hobbies
    - ☐ Starting your own business
    - ☐ Other___

- ☐ Plan so you will have about 80 percent of your pre-retirement income as a retiree, if you intend to continue your present lifestyle.

CHECK YOUR ASSETS
- ☐ Do you own a home?
- ☐ Will your mortgage be paid off by the time you're ready to retire?
- ☐ Will the size of your house, or the taxes, cause a financial strain?
- ☐ Look **now** into all these sources of income. What sources will, or could, become available to you as a retiree?
    - ☐ Social Security
    - ☐ Corporate Pension Plans
    - ☐ Employee Savings Plans
    - ☐ Supplementary Retirement Plans:
        - ☐ IRAs
        - ☐ Keogh Plans
    - ☐ Life Insurance Endowments and Annuities
    - ☐ Other_____
- ☐ Choose wisely from among the income sources available and begin retirement preparations now.

CHAPTER 23

PARTING ADVICE

Hardly anyone would make a major purchase—such as buying a television set or a home computer or a microwave oven—without giving it some thought. Most of us shop around, compare prices and options and features and quality, and then select the model which best fits our needs—and our pocketbooks—from the dealer in whom we have the most confidence. When the issue is our financial security, selection of the appropriate product or service is of far greater importance.

It is hardly necessary to repeat that a careful and informed choice is the keystone of successful investing, saving and retirement planning. Most people recognize the need to devote thought and study to managing their finances. As consumers of insurance, all of us need to exercise our best shopping skills. Unfortunately, the same folks who spend weeks looking for the perfect car sometimes spend hardly any time at all seeking out the best insurance coverage. And they may give no thought at all to making insurance a consideration in selecting the car! Nevertheless, the kind of car you drive can have an impact on your auto insurance rates. The way to choose the right car—and the right insurance—is to know about all the available options.

Bernie Hoffman, for instance, asked only half the right questions. Being interested in cars, he asked all about gas mileage on a foreign make. Deciding he could save significantly on gas, he decided to buy a small two-seater model. What he hadn't asked about was insurance cost for that car. There turned out to be a sports car surcharge, which more than eliminated his projected gas savings.

Economists generally look on an informed consumer as a basic part of the free enterprise system. It is assumed that consumers in a competitive marketplace are aware of the many products available to them and that they have the information and wisdom to choose intelligently from among them. This implies that the companies that sell products and services—including financial products and services—are obligated to provide information about them. And that consumers have a right to that information. But becoming informed isn't automatic, or effortless. It requires some study, and sometimes a little legwork.

How does one go about becoming well informed about products related to financial security? Reading this book is one good step. Here are some others.

Plan Your Program

Analyze your goals and your needs whether for insurance, savings, investment, retirement or estate building. Survey the assets you have to protect, and the various means of protecting them that are available to you. Draw up minimum, maximum and in-between programs, balancing what you can afford in terms of cost and what you want in the way of returns. Don't let your programs grow haphazardly. If they already have done so to a certain extent, draw up a long-range plan and see how your existing programs fit. If your analysis shows major goals unmet, or certain costs duplicated unnecessarily, be prepared to change your programs. You can evaluate a financial product or service only in relation to what it will do for you in terms of your overall goals and needs. A suggested approach might be to list your objectives on paper. For example:

1. Income.
2. Savings and investments.

3. Property (e.g., home, car, personal possessions, other property).
4. Education (yours or your children's).
5. Retirement.

Then, under each heading, list the various possible means available to you for achieving the objective. For instance, under savings and investments, you might list real estate, stocks and bonds, mutual funds, bank accounts, and cash value of life insurance. For each item, put down what your minimum needs are, and what goals you can reasonably hope to achieve. Against these, set down what you already have, and what gaps need to be filled. This process will give you a picture of both where you are and where you want to be.

Invest Some Time and Energy

You can expect a substantial return in financial security if you take the time to look carefully at that new disability policy being offered you and note how—or whether—it coordinates with your health insurance plan at work. Or maybe you need to locate your renter's policy and look it over, just to remind yourself that you've got liability coverage, in case a guest slips on your bearskin rug.

Spending a little time and money while you're still considering the purchase can pay even bigger dividends . Suppose you've saved $10,000. Here come the decisions. Should you place it in a money market fund or a certificate of deposit? Or should you take $2,000 off the top and start that IRA you've been thinking about? These options are worth plenty of consideration. And they're worth some homework, before you decide. Otherwise, you're likely to act on incomplete information—or worse, on impulse. Take the case of Norma Stiller, who was unexpectedly left several thousand dollars on the death of a relative. Not having given investing any previous thought, she acted impulsively on a tip from a friend and bought a "penny" stock, which was rising rapidly in market value. Too late, she found that the stock had gone up, and her equally naive friend had heard about it because everyone on Wall Street had heard about it earlier, and the stock was enjoying a flurry.

After Norma bought the stock at its high point, it promptly went into a decline as the Wall Street "profit takers" sold out. Ultimately, she sold out too—wiser, but poorer.

When you shop for insurance, you'll need to have a handle on what it is you want to insure. How much at current prices would it cost to rebuild your house if it were destroyed in a fire or a tornado? Having an up-to-date appraisal or estimate would help. And what about your personal belongings? Having an accurate, up-to-date inventory with you when you talk with an agent or company representative could be a big advantage.

Suppose you've just traded your eight-year-old car for an almost-new model. Before driving your "find" off the lot, better review your auto insurance policy. Dropping your collision coverage a couple of years ago probably was a great idea—then. But this newer car is another matter altogether. And maybe you should reassess your liability coverage while you're at it. Perhaps your income has grown and you have more assets to protect.

Your agent or company representative will be happy to discuss all these matters, but you'll be way ahead if you do your homework before talking to either one. That way, you'll know what you're discussing when you're making decisions about your financial security. Your accuracy in providing information about your assets and liabilities also will enable a professional insurance adviser to make better recommendations about the kinds and amounts of coverages most suitable for your needs.

Seek Out Professional Advice

When you're setting out to find help on questions of financial security, you'll want to know you're speaking with a "pro." When you move to a new town, or acquire a new need for coverage, your search for an insurance adviser should be as serious as your interest in finding a good doctor or lawyer.

There are four general types of professional insurance advisers.

- An independent agent is a self-employed businessperson who usually represents two or more insurance companies in a sales and service capacity and is paid on a commission basis.

- The exclusive agent represents only one company, usually on a commission basis.
- The direct writer representative is the salaried or commissioned employee of a single company which sells through its own employees or by mail.
- The broker is a market specialist who represents buyers of property and liability insurance and who deals with either agents or companies in arranging coverage for the customer.

You may wish to talk over your insurance needs with several of these advisers, as a means of comparing costs and coverages and to locate the policy or combination of policies best suited to your particular situation. Some auto insurance companies specialize, for example, in products tailored to the insurance needs of older drivers or non-drinkers. Others cater to servicemen or to collectors of antique cars. Some provide homeowners insurance exclusively to owners of single-family houses, others to renters or condominium owners.

Ask friends or relatives about agents and companies from whom they've had good service. If you're looking for an independent agent, you'll want to find out whether he or she is an active member of an organization such as the Independent Insurance Agents of America or the Professional Insurance Agents. These groups conduct special educational programs for their members and help in keeping them up-to-date on new products and developments in the insurance marketplace.

If your insurance needs involve your business as well as your personal insurance, you may want to deal with a broker. A broker, rather than representing one or several insurance companies, is a professional who shops the entire market in order to put together an insurance program for you. Most are oriented to commercial accounts. Your main question in selecting a broker would be, "How much time and attention will he give my account?" Many independent agents are also licensed as brokers.

The better informed your insurance adviser is, the better equipped he or she is to design a package of insurance suited to your individual needs. The attainment of the title of CPCU (Certified Property/Casualty Underwriter) or CLU (Certified Life Underwriter) indicates the individual is a serious professional who has completed an intensive course of study and passed rigorous examinations. However, years of experience in handling policyholders' needs and problems can be an equally important qualification.

It's best to deal with a full-time professional rather than with a part-time insurance salesman who has another business and sells insurance on the side. It stands to reason the full-timer is going to have more detailed information about his products, as well as more time and interest to give your account.

In buying life insurance, it's better to seek out a reputable company or agent, rather than wait to be sold a policy by the first person who comes to your door. You also should shop around, and compare policies and values, before making your decision. That way, your chances are better of getting your own needs met, rather than the salesman's. It's too bad most people remain unaware of the place of life insurance in their planning until a salesman points it out to them. You'll do better as an aggressive insurance "buyer" rather than a passive "prospect" whose needs may or may not coincide with the benefits of a proffered policy.

Aside from paper credentials, how can you tell if an insurance adviser is the right one for you? By talking to him or her. The "right" person will be in tune with you, will be interested in your problems, will take the time to explain things to you, will know or find the answers to your questions, and will inspire your trust and confidence.

In talking to an adviser, first state your needs, then ask each prospective agent or company representative what kind and amount of coverage he or she would recommend for you. To get a good recommendation, it'll be vital for you to provide accurate answers to the questions asked you—questions about your house, car, children and also about your personal finances, assets, liabilities.

If an agent recommends a policy from an insurance company you haven't heard of, you may want to look the company up in Best's Insurance Reports, which you can find in many libraries. There, in addition to information about the company's location, officers and recent financial records, you'll find an overall letter rating. Give first consideration to companies with ratings of A or A+. Remember that the quality and service of that company will be of paramount importance when you have a claim. As is true with other products, the policy with the lowest price may not always be the best bargain.

Insurance is one of several important factors which play a part in achieving and maintaining financial security. And a professional insurance person is just one of the "pros" who should make up your team of financial advisers.

If you're a novice at dealing with the stock market and need advice on what stocks to buy and when to buy them, you may need to locate a professional who is associated with a full-service brokerage house. Again, just as with insurance agents, it's important to talk with several prospects. And again, you must do a bit of homework in advance. For example, you should have in mind how much you're willing to invest and whether you're looking for quick returns or long-term growth. And don't forget you'll need to understand the recommendations your broker will be making. If you don't, will he or she be willing to explain? What about the size of your portfolio? Ask during your interviews whether it's about average in terms of those handled by the firm. If it's too small, you may not get the attention you'll need. If you're a small investor, deal with a firm which caters to the small investor. If you have substantial assets, a bank trust officer or an investment counselor who will take a personal interest in your account may be your best bet. Before choosing an adviser, check on past performance. Investment returns are published by banks and investment firms. Ask to see them.

Your banker plays a significant role in your financial security, so you'll want to cast a real "pro" in that part, too. Are you thinking of having your checking and savings accounts and of opening an IRA all in the same bank? It's worth your while to do some comparing. Does this institution offer a wide range of services? For example, is it likely to be a source of mortgage money for that condominium you want? Will this institution provide an officer who is willing to discuss your interest in IRAs and to answer your questions without giving you a "brush-off"? Is there a trust department that will advise you on investments, on government securities, on municipal bonds, on current yields? Find out what the bank is paying on its deposits—its certificates of deposit (CDs), IRAs, passbook savings accounts and other types of accounts that interest you. If you find bank personnel brusque or impersonal, perhaps a smaller bank or smaller branch is for you.

Ask Questions

Ask about anything you don't understand—before or after you insure, invest or commit your money. After all, it's your financial security that's at stake. You have the right, and also the responsibility to yourself, to know about developments that may affect your future.

Be sure you understand what your insurance policies provide—and what they don't. Take the time to look over your policies. You're not expected to be an expert. But you have an agent or company representative who is, and one of the services you pay for when you purchase a policy is the benefit of that person's expertise. So don't hesitate to give your agent or company a call when you're not sure about your coverage, when you add to your property, or when you want to ask any question at all.

The Wilsons had a homeowners policy for years but were unaware that liability coverage is a part of that package of insurance. So when Timmy from down the block fell out of their Susan's treehouse and broke his arm, they didn't realize they could have called on their homeowners policy to help pay his medical expenses. Instead, they volunteered to pay the bills out of pocket. A costly mistake. Not the way to get the best value from insurance dollars.

Another mistake, ranking right up there with not knowing you're covered, is making the assumption you are covered—without asking. Either way, you're not making the most of your insurance and worse, you may be placing your financial security on the line.

Many of the citizens of Fort Wayne, Ind., doubtless were shocked when their city was covered with water during the terrible flood that made news in the spring of 1982. Some of them may have experienced an equally devastating blow as they asked—too late—whether their homeowners insurance covered their losses. It didn't. No homeowners insurance does, because flooding occurs only in certain areas.

But in Fort Wayne, flood insurance was vital. And, for the homeowners who had asked their agents about flood insurance coverage and had bought it through a special federal program, there was good news. Their financial security, at least the part of it that was vested in their homes and belongings, was largely intact.

Your financial security is worth asking about. If you're thinking of taking early retirement but are not sure whether you can afford it, why not contact the local Social Security representative and ask? Find out whether you could take that part-time job your company is offering and what impact it might have on your benefits.

Or, maybe your "little Charley" (who's now 6'3" and weighs 185) is going to be ready for college next year. Feel free to phone or visit the professionals to whom you've entrusted your financial security. Ask whether you can borrow money on your life insurance policy or use the equity you've built up in your home to finance his education. Some banks have offered programs that give you cash for your home while allowing you to continue living in it for your lifetime. Maybe an education now would be worth more to your son than willing him your house when you die. Explore the options.

Look for Money-Saving Ideas

Concentrate on taking advantage of deductibles and discounts, and avoiding duplications.

What if you're just looking for new ways to trim expenses so that you can foot Charley's tuition bills out of current income? Ask your insurance agent about increasing your deductible and find out exactly what that would do to decrease the premiums on your car and home coverages.

Take the largest deductible you can afford, provided you really do save on the premium reduction. That is, decide how large a loss you could absorb on your own without suffering severe financial hardship. Imagine the following scenario as an example. Suppose you're late for work one morning and you back your car out hurriedly, not remembering first to open the garage door. (Don't laugh—it has happened.) There's no damage to the car, but the door is splintered, and necessary repairs will come to $195. Can you afford to pay for this loss and still take care of other financial obligations, such as a mortgage payment, car payment, insurance premiums, food and school-related expenses? If you can, then you should have at least a $200 or $250 or maybe even a $500 deductible. If you can't, better consider insuring against all losses above $100. But the higher deductible will usually save you money each time you pay your premium.

A costly mistake consumers make too frequently is using insurance to pay for small losses. Remember, the proper use of insurance is to pay for large losses, those which would be unaffordable or cause real financial hardship if you had to absorb them yourself.

Is Charley going to school at least 100 miles from home? If so, you may be eligible for a discount since he'll be driving the family car only infrequently. If he's an honor student, that's good for another discount.

While you're at it, ask about any other discounts that might apply to you. If you've joined a car pool, or taken a course in defensive driving, or installed special anti-theft devices in your car, you may be eligible for a discount. It's certainly worth discussing with your insurance adviser.

Avoiding duplications is another money-saving measure. If, for example, your group health insurance policy at work gives you broad coverage, you won't need coverages for specific diseases or other special medical policies. If you have an up-to-date homeowners policy, you generally won't need separate fire or theft or liability insurance, since these risks already are covered (unless you're engaged in a business at home).

Combine Your Coverages

Package policies come cheaper than separate coverages. It's cheaper to insure all your cars under the same policy, and it also pays to carry all or as much of your insurance as possible with the same company. A Midwestern agent tells of a client, a doctor, who persisted for years in insuring his practice, and also several other businesses he had an interest in, and his personal property, under separate policies with different companies. The agent took the initiative to put together a complete package for him with a single insurer, and showed him how he could make a significant saving. The doctor accepted, and admitted that his past ways were his own kind of "insurance" against cancellation or the bankruptcy of one of the companies. But the doctor wasn't as expert on insurance as he was on medicine; the agent pointed out to him that the state's insurance laws protected him against arbitrary cancellation and also against insurer insolvency through a state guaranty fund, contributed to by insurers and designed to assure payments due to policyholders.

Consumers who take the initiative in asking questions and in looking for ways to save are likely to benefit more from their investments and insurance than those who take a passive approach. And they also are likely to be progressing more rapidly toward the goal of financial security.

Reassess Annually

It's important to re-evaluate the various elements of your financial security program at least once a year. The principle of periodic re-evaluations always has been valid, but in the volatile economic climate of the past few years, it has become vital.

Consider the array of financial products available. Some of them did not even exist until recently and others have undergone marked changes to meet new consumer needs—and thus require new consideration. Money market funds, interest on checking accounts, Keogh Plans, IRAs, umbrella liability insurance, single-limit auto liability, replacement cost coverage for household belongings—these are just a few examples.

The impact of inflation also has made an annual reassessment of investments and insurance a necessity.

Look at the record. From 1970 to 1980, on average:

- The cost of buying a home increased 115 percent (which means that a house which cost $30,000 in 1970 would have been worth $64,500 in 1980);
- House maintenance and repair costs rose 130 percent;

- Private residential construction costs skyrocketed by 156 percent.

Since, for most people, the home represents a major investment—and thus a highly significant part of financial security—it's easy to see that these figures deserve careful and frequent scrutiny. Policy renewal time provides an opportunity to review the coverages of your home insurance and the possible need to update them.

Both the impact of economic inflation and the trend toward increasingly higher court awards in personal injury cases make it imperative that you review your liability coverages regularly. Million-dollar and even multi-million-dollar verdicts in serious cases are not at all uncommon. Thus it is only prudent to consider carrying high liability limits—and to consider umbrella liability coverage—especially since the extra cost is relatively small.

Changes in your life or your lifestyle offer still more reasons for a yearly look at your insurance. Have you changed jobs, moved to a new city, decided to drive that old car awhile longer? If you've done any of these things—or if you've had a baby or installed a sauna—your earlier insurance decisions now may be outdated. Phone your agent or company representative to talk over any significant changes in your life or your household and get some professional advice as to whether your policies need updating.

An Ohio couple, who had two cars on the same policy, split up. Each took one of the cars. Shortly afterward, the wife called the insurance agent and told him to cancel the husband's coverage, saying, "We no longer have that car." Just to be sure, the agent asked her to send him a written request for cancellation. The letter never came, and at renewal time, the agent got in touch with the husband, who explained the situation and asked for a renewal policy. Except for the agent's alertness to marital "vibes," the man could have been caught without auto insurance. The moral is that it's best to put your insurance representative on your list of confidants in such family matters.

Keep in mind that updating your insurance coverage to meet your changing needs does not necessarily mean increasing either the coverage or the premium. If you've added a new swimming pool to your backyard "estate," you may or may not need to increase your liability coverage. On the other hand, if you have a new driver in your household who'll be using the family car—like the daughter who just got her license—you must have her added to your policy.

If you've added a new baby, this is an excellent time to reassess your entire financial plan, including your life insurance, savings programs and investments.

If you've become a single parent, your annual reassessment might indicate a need to change the beneficiary on your life insurance policy or perhaps to consider additional coverage for the benefit of your child's financial security. Also, ask yourself whether it is possible or advisable to include him or her in your health and hospitalization program at your place of work.

On the other hand, if your children have grown and completed their educations, you may be able to drop some of your life insurance, or switch the investment to your retirement plan.

Be careful about switching insurance companies to get a lower rate. It may be that the other company has not yet declared its rate increase, and is about to do so, making your gain only temporary.

Don't Buy Insurance You Don't Need

If you are still single, or if you have become single again, you may have little need for life insurance unless you wish to use it as a savings vehicle. If you are taking a plane trip, and you already have sufficient life insurance, you may decide you don't need to buy additional flight insurance—unless it does something for you psychologically, gives you some comfort if you're a nervous airplane passenger, that is worth the extra premium to you. If you rent a car occasionally, you don't need to buy extra auto insurance from the renter if you have your own auto collision coverage with a deductible you're comfortable with. In considering any insurance purchase, review your existing policies to see if you're already covered. And weigh the risk. Is the potential loss one you could afford to absorb yourself? If so, you don't need, and probably shouldn't buy, insurance. The one coverage you shouldn't skimp on is liability insurance, because of the potential for lifetime financial ruin that lurks in personal liability.

In the category of insurance you definitely don't need is duplicate insurance or overinsurance taken out in the expectation of a windfall in the event of a claim. Many companies protect themselves against this by wording, or a clause in their policies, which prevents you from collecting on claims in any amount greater than your loss.

Keep an Inventory

Compiling an inventory of your furniture and personal belongings for insurance purposes has been recommended. But you should also make a different kind of inventory—of all the documents pertaining to your program of financial security. Not the documents themselves, but a listing of those documents.

You could prepare a card file, for example, with information about each insurance policy, savings account or investment. A typical insurance inventory card might look something like this:

Car Insurance

Agency:	Jones & Son 69 W. Main Street Clarkeville, Mass. 07320 (208) 777-1234
Company:	XYZ Insurance Co., Inc. 2 Sherman Square Portland, Maine. 23712
Policy No.:	237 W 6294X
Coverages:	Property damage liability Bodily injury liability Collision Comprehensive Medical Payments Uninsured Motorists
Amounts:	100/300/50 thousand $200 deductible

Car(s) Listed: 1979 Buick
                 1983 Datsun

Drivers Listed: John G. Doe
                 Mary F. Doe
                 Carla T. Doe

Generally, all your cards would contain these kinds of data:
- —A category or title (e.g., life insurance, savings account) in a prominent place.
- —A specific company, agency, bank or brokerage firm (including name, address, telephone number).
- —Address of the home office or branch office where your account is located, and the policy or account number.
- —As much specific information about amounts of money or coverages as possible.

Your financial security inventory could take the form of a file folder with a sheet of paper for each entry. If you wish, you might even include a photocopy or photostat of each document.

Making an inventory is not terribly time-consuming. An afternoon or an evening with the project can give you a real sense of satisfaction and add to your confidence that your life is under control. Once you've done the job, it's a quick and easy matter to replace or add one new card or sheet to your file as you make new financial decisions.

There's no right or wrong way to make an inventory. The important thing is to do it—now! Just make it your aim to include as much information as possible The more complete your inventory, the more valuable it will be when you need it.

Keep a duplicate copy of your inventory in a safe place away from home, preferably in a safe deposit box. It's a good idea to keep the original documents there, too. That way, if you have a fire, the only copies of your insurance policies, your savings passbook, and the deed to your home—your financial security—won't go up in smoke.

Know Your Resources

When you have a question about your hospitalization plan or a problem understanding a letter from your banker or a dispute over the interest on your certificate of deposit, it's important you know the courses of action, the resources, the expert advice available to resolve the matter. With insurance questions, a look at your policy is a first step in finding answers. Checking with an expert is next. Your own insurance representative or, in some cases, the staff member who handles your insurance program at work should be your primary source of information about what is covered and how much is covered. And those same experts are the people to call or see when you need to file a claim.

Learn How—and When—to File a Claim

Let's suppose those crumbs in your old toaster start a fire in your kitchen. You're able to put it out right away because you keep a small fire extinguisher in the hall closet, but still, there's smoke damage to your brand new wallpaper, and your kitchen counter is ruined. Your local insurance agent or company representative is the first person to get in touch with. He or she may ask you to fill out and submit a written form giving details of the accident or loss. A claim representative may ask you to make a list of all your damaged property, and propose a settlement.

Very simple. But what if you don't think the settlement is adequate as described in your policy?

Your agent—the person from whom you purchased your insurance—is still your best source of advice and recourse. Go back or phone again. Explain that you are concerned.

Once you are sure your agent understands your problem, and if you remain dissatisfied with the settlement, or with his or her response, call the home office of your insurance company and ask for someone in the consumer affairs or customer relations department. If you need help in locating your company or are having trouble getting through to the appropriate person, call the toll-free hotline in my office at the Insurance Information Institute (800-221-4954) for assistance. In New York State, call 212-669-9200, collect.

If your problem still has not been resolved, then call or write your state insurance department.

To manage your personal risks and protect your assets, it may help you to realize that (1) you don't have to be an expert, (2) you do have to do some planning, and (3) through your agent or insurance representative you have resources available to help.

There are no magic formulas. Your financial security hinges on personal decisions only you can make, and actions only you can take. If this book helps you to make better informed decisions, its purpose will be accomplished.

APPENDICES

APPENDIX A

HOW TO READ YOUR INSURANCE POLICY

What is that printed piece of paper called an insurance policy?

If it seems rather legal-looking, it's because it's just that—a legal document. An insurance policy is a contract. Historically, the language used in insurance policies evolved largely out of court cases and decisions. Policies therefore have been designed to be precise, in a way that will stand up in court, about the rights and responsibilities of the parties to the contract—you, and the insurance company. More recently, insurance companies have recognized the need for policies to be clear and understandable to consumers as well as to lawyers, and have devoted considerable effort to simplifying the language and format of many policies.

The responsibilities of the insurance company under the policy contract are to pay your legitimate claims, represent you in court, and perform related services as stated or implied. As a policyholder, your responsibilities are few but important. Principally, they are to pay your premium, to give honest and accurate information, to follow procedures outlined in the policy when making a claim, and to abide by other conditions as described in the policy.

The types of insurance fall into two main categories: personal lines and commercial lines. Since this book is intended for you, the individual consumer, it is concerned only with the personal lines—those kinds of policies designed for your protection as an individual person—and not with those designed for a business organization or an institution.

Now let's take a closer look at that document called an insurance policy. What's in it? In many cases, there is a great deal more in it than there used to be. When the late John D. Rockefeller, Sr., bought insurance for his Westchester County, N.Y., estate, Pocantico Hills, he had to buy a half-dozen different policies—a fire insurance policy, a policy against windstorm, a burglary policy, various kinds of liability coverages, and policies against any other risks that he might be concerned about. Today, his heirs can get all of those different kinds of protection wrapped up in a single "package" policy.

Most auto insurance and homeowners policies now are packages containing several different kinds of coverages. A few companies have experimented with a policy combining coverages for both home and car in a single package. Perhaps because most people buy their homes and their cars separately and therefore tend to insure them separately, this super-package has not yet come into widespread use.

But there are advantages to you in using the package policies. First, there is usually a significant cost saving, since combining coverages lowers overhead costs for insurers and enables them to charge you a lower premium than for coverages bought separately. Second, there is the benefit of convenience to you, since you have only one document to keep track of, one company to deal with, and one claim to make in case of a multiple loss. While homes and cars usually are not damaged in the same accident, it can happen—in a fire, for instance.

An Omaha, Neb., agent recalls a customer who came to him to have his house, his car, his pickup truck, and his boat and trailer all insured together under a package policy. A short time later, a tornado came through the town, wrecking property of all kinds indiscriminately. Most homeowners hit by the tornado had to deal with two or more adjusters to settle claims on their homes, their cars and their other property. But the man with the package policy, whose property was damaged, was able to wrap up a complete settlement quickly, with a minimum of fuss, by dealing with only one adjuster.

The exact premium saving on a package policy compared to separate coverages would vary depending on the company and the coverages involved. But if you're shopping for insurance, you can easily determine the difference—which will be sizable—by asking your agent or sales representative for comparative price quotations.

That difference is a main reason why premium volume for the homeowners package policy multiplied by more than 45 times between 1957, when the insurance industry started keeping track, and 1981. Developed by insurance companies as a competitive marketing tool, the homeowners and automobile package policies have become the norm, saving consumers billions of premium dollars.

The different home insurance coverages still are sold independently of each other—not all homes are eligible for packaged coverage, and many people still find a need to supplement what's in the package. But by and large, package policies are the "in" thing today. Specific coverages contained in the auto and homeowners packages are discussed in earlier chapters on those subjects.

Your protection under a package policy, though, covers two main items: your property, and your liability for damages to others. Property insurance is very simple—it reimburses you for loss of or damage to things that you own. Liability is a bit more complicated, and it's worth a little discussion.

Liability and the Law

Liability law traces its roots back to ancient concepts of retribution for wrongs done, further developed in English common law under the concept of torts. A tort is a wrongful or negligent act, resulting in injury or damage to someone's person or property, for which the law permits the other party to sue you for recompense in civil court (it is not a criminal charge). The outcome depends on a determination of who was "at fault"—responsible for causing the damages.

Basically, liability insurance protects you in two ways:

1. If someone institutes a claim or a lawsuit against you, your insurance company will take over, with the aim of either settling the case out of court or providing for your legal defense. In doing so the insurer assumes control, and you'll be expected to provide whatever assistance and cooperation may be required. (If you undertake to settle a claim against you on your own, your insurer can and probably will wash its hands of the matter and you'll find yourself shouldering the expense.)
2. If your insurer agrees, or a court decides, that you are legally liable, the company will pay the damages assessed against you up to the limits of coverage specified in your policy. ("You," in this context, means anyone who is covered by your policy.)

Policies providing personal liability insurance also require the insurance company to pay such expenses as:

- Premiums on certain kinds of bonds which may be required (for example, a bail bond);
- Interest that might accrue on a judgment before it is finally paid;
- A flat daily payment, usually up to $25 or $50, for loss of earnings if you attend hearings or trials at the company's request, and
- Other reasonable expense incurred at the company's request.

A finding of legal liability means that you, somehow, in the eyes of the law, have been guilty of negligence. You may consider yourself a careful person, but negligence can take many forms—from the failure to make Johnny put his skates away where people can't trip over them to smashing into another vehicle with your car because you failed to see an icy patch on the road. The fact that the damage may have been unintentional may protect you from criminal charges, but it won't protect you from a civil liability suit. For example, there was the case of an Ohio woman who was invited to visit a friend for a winter morning kaffeeklatch. On arriving, she slipped on the icy steps and broke a front tooth. Although medical bills were only $500, she sued for and collected $4,000, for her friend's negligence resulting in her personal injury.

A more serious, but still typical, case was that of the Nebraska woman whose young son was playing with another boy in her yard. Their game was to throw clumps of dirt at each other. A hard clod knocked out the other little boy's eye. The injured boy's parents sued, and collected $10,000 in a liability judgment against the other parent.

Contributory and Comparative Negligence

Often a court will find that both parties to a lawsuit were partly at fault. What happens then depends on whether the state has a contributory or comparative type of negligence law.

For example, suppose you run a red light and crash into a car emerging from a side street. Clearly your fault, but suppose the side street was a one-way thoroughfare—the other way than the suing driver was going. Under the socalled contributory negligence laws in effect in some states, if the person who is doing the suing is found to be partly responsible—even to a small degree—that constitutes grounds for dismissing the case.

A growing number of states take a less hard-nosed approach. They have what are called comparative negligence laws, which say you may be liable for damages in proportion to your share of the fault. In the red light-wrong way illustration above, assuming that the total damages to the other person were $10,000, and the court decided the accident was 75 percent your fault and 25 percent the other person's fault, you would be liable for $7,500.

Special Responsibility to Children

Although the law holds you responsible for exercising the degree of care toward anyone that would be expected of a reasonable and prudent person, the courts have held that you owe a special degree of care for children. For example, under the "attractive nuisance" doctrine to which most courts subscribe, it can be a costly mistake to put temptation in the way of young children—even trespassing young children—whose immaturity and curiosity may lead them into situations where they can get hurt or cause harm.

If the temptation of your unfenced swimming pool or your car left unoccupied with the engine running were to prove irresistible to a youngster, you could find yourself facing an expensive lawsuit should the child drown in the pool, or put the car in gear and hit someone.

How Much Protection?

The liability insurance most people carry, whether through a package policy or a separate personal liability policy, will meet their needs in most circumstances. But standard policies have a dollar ceiling (called limits) on how much they will pay, and there are some situations, such as libel and slander, that they don't cover.

To help people protect themselves against possibly ruinous losses in situations that go beyond the coverages and limits of a standard policy, insurance companies offer a special kind of policy called umbrella liability. Umbrella liability pays for losses above the limits of your primary policy.

Limits of umbrella liability policies are much higher than in standard policies, commonly ranging up to $1 million or $5 million. Most umbrella policies come with a deductible, and most insurers will sell you an umbrella policy only if you have a certain minimum of coverage in your primary policies, say $50,000 in a homeowners policy and $300,000 in an auto policy. That amount becomes the starting point at which your umbrella "excess" coverage takes over. Since umbrella coverage would rarely come into play, the coverage is relatively inexpensive and is a good buy especially for anyone with substantial assets to protect or who might be especially vulnerable to lawsuits and costly judgments.

How an Insurance Policy Is Put Together

Insurance policies lose much of their mystery when you realize that most policies covering homes or cars have similar elements.

How to Read Your Insurance Policy

The typical property/liability policy has four parts:

- The declarations page. Names the policyholders, describes the property or the liability to be covered and states the type of coverage and the maximum dollar limit that the insurance company will pay for a claim.
- The insuring agreement. Sets forth the policyholder's and the insurer's responsibilities.
- The conditions of the policy. Spells out details of the coverage and what is required of the insured and the insurer in the event of a loss resulting in a claim.
- The exclusions. Details the types of property and the types of losses which are not covered, and conditions under which coverage doesn't apply.

In the package policies, property and liability coverages are dealt with separately.

Some of the conditions common to many property/liability policies include:

- Concealment, fraud. If you willfully conceal or misrepresent any facts in applying for the policy or in filing a claim, the insurance company may legally declare the policy void and refuse to pay any loss.
- Perils not included. This goes a step beyond other exclusions contained in the policy by getting very specific as to certain things not covered, such as losses resulting from war or insurrection.
- Waivers. Provisions of the policy may not be changed or waived except in writing by the company.
- Cancellation. You may cancel your policy at any time.

  The insurance company may cancel it, with advance notice to you, under certain conditions. The remainder of the premium, on a pro rata basis, will be returned to you if the insurer cancels. A charge for administrative costs will be made if you cancel.
- Mortgage interest. Since some property covered under a policy may be mortgaged, the lender has a vested interest in the insurance on the property, which the insurance company recognizes. This could mean, in the event of total loss by fire, for example, that the lender could elect to liquidate the mortgage. You would receive the remainder of the insurance proceeds.

- Duplication of coverage. This clause prevents profiteering from losses by providing that if more than one policy is in force for the property, each company will pay a proportionate share of any loss.
- Requirements in case of a loss. The policyholder is held responsible for, among other things, prompt notification to the company; protection of the property from further loss; submission of proof of the loss; and making available to the insurance company necessary documents and financial records.
- Appraisal. In the event you and your insurance company disagree on the amount of a loss, this clause outlines procedures for selecting and paying competent independent appraisers to determine a settlement.
- Company's options. Where a loss involves property damage, the insurer may take all or part of the damaged property after payment at an agreed value; repair, rebuild or replace it with material of a similar kind and utility; or settle in cash.
- Subrogation. After paying a claim to a policyholder, the insurance company has the right to recover the amount of the payment from a third party who was responsible for the loss. The policyholder waives the right to collect twice—from the other party as well as the insurer.

How to Read Your Policy

Every policy contains carefully worded definitions of certain words and phrases as used in the document. In the newer simplified policies, the words and phrases which are defined usually are printed in bold type or italicized wherever they appear.

Understanding your policy is not really as difficult as it might seem. It's in your own interest to read it. Even if you don't have the time or inclination to read the whole thing right away, it's important that you check the declarations page—even before you pay the premium—to make sure that the policy covers what you want it to cover and in the proper amounts. If everything isn't presented to your satisfaction, including such details as your name and address, get in touch with your agent, broker, or company immediately and have it straightened out.

Keep the various parts of the policy in mind as you read from one point to the next. After you read what's covered, what your responsibilities and those of the company are, and the conditions of the policy, check the exclusions and other conditions. They are there to clarify exactly how the policy is intended to work.

Take this statement in a simplified auto policy:

"We will pay damages for bodily injury or property damage for which any *covered person* becomes legally responsible because of an auto accident."

The phrase "covered person" is in bold type or italics. That's a reminder to check the definition. In this case, "covered person" can be any one of several people in a family.

The policy will generally track with common sense. Is the sky the limit as to when and how much the company will pay? Of course not. Check the exclusions. One of the things you'll find there is a stern warning that the company has no responsibility for intentionally caused bodily injury or property damage.

Finally, even though the policy's dollar limits are stated on the declarations page, the insurer usually puts it into words something like this in a paragraph following the exclusions: "The limit of liability shown in the Declarations for this coverage is our maximum limit of liability for all damages resulting from any one auto accident."

Insurance companies make an effort to be precise, even in the simplified policies. If, after reading the entire policy, you still have questions or feel there are things you don't understand, don't hesitate to call your agent or company with the questions. You've just made an important purchase; it's your right to know exactly what you've bought.

APPENDIX B

YOUR CHOICE OF HOME- OWNERS POLICIES

As the owner of your own home, you usually have a choice of several policies, although realistically a vast majority of homeowners narrow their choices to two. The one you should select depends on what kinds of losses you want to protect yourself against. The various forms (numbered HO-1 to HO-8) and the coverage they provide are shown in the accompanying chart.

The homeowners packages cover homes valued at $15,000 or more.

The various forms differ in the number of "perils" covered, and in whether the coverage is on a "named perils" basis—that is, with the covered perils specifically listed—or on an "all risks" basis with all perils covered except those specifically excluded.

The broader the coverage, usually the higher the premium. But the extra coverage may be well worth the extra cost to you. The best way to economize on your homeowners insurance is to weigh the specific protections you get against the extra premium, and choose the policy that gives you all the protection you need, but doesn't provide coverage you don't need or feel you can get along without.

You should go over the options in detail with your agent or company representative at the time you are making your purchase. It's an important decision, so don't rush it at that point. Package policies

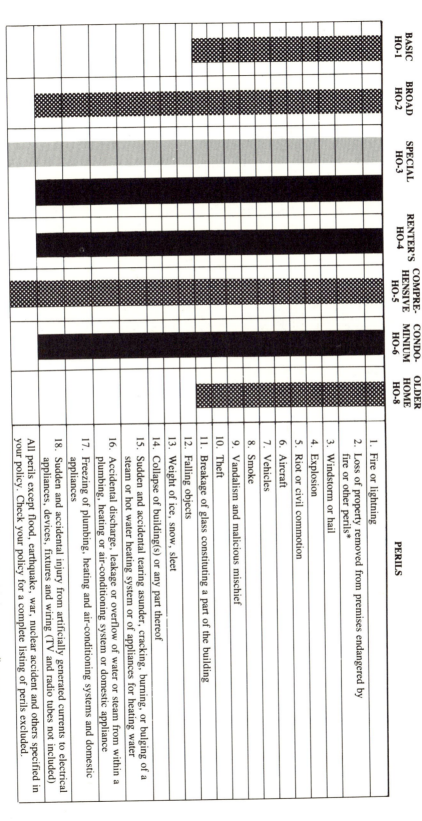

PERILS AGAINST WHICH PROPERTIES ARE INSURED UNDER THE VARIOUS HOMEOWNERS POLICIES

PERILS	BASIC HO-1	BROAD HO-2	SPECIAL HO-3	RENTER'S HO-4	COMPRE-HENSIVE HO-5	CONDO-MINIUM HO-6	OLDER HOME HO-8
1. Fire or lightning	■	■	▨	▨	■	▨	■
2. Loss of property removed from premises endangered by fire or other perils*	■	■	▨	▨	■	▨	■
3. Windstorm or hail	■	■	▨	▨	■	▨	■
4. Explosion	■	■	▨	▨	■	▨	■
5. Riot or civil commotion	■	■	▨	▨	■	▨	■
6. Aircraft	■	■	▨	▨	■	▨	■
7. Vehicles	■	■	▨	▨	■	▨	■
8. Smoke	■	■	▨	▨	■	▨	■
9. Vandalism and malicious mischief	■	■	▨	▨	■	▨	■
10. Theft		■	▨	▨	■	▨	
11. Breakage of glass constituting a part of the building		■	▨	▨	■	▨	
12. Falling objects		■	▨	▨	■	▨	
13. Weight of ice, snow, sleet		■	▨	▨	■	▨	
14. Collapse of building(s) or any part thereof		■	▨	▨	■	▨	
15. Sudden and accidental tearing asunder, cracking, burning, or bulging of a steam or hot water heating system or of appliances for heating water		■	▨		■		
16. Accidental discharge, leakage or overflow of water or steam from within a plumbing, heating or air-conditioning system or domestic appliance		■	▨	▨	■	▨	
17. Freezing of plumbing, heating and air-conditioning systems and domestic appliances		■	▨	▨	■	▨	
18. Sudden and accidental injury from artificially generated currents to electrical appliances, devices, fixtures and wiring (TV and radio tubes not included)		■	▨	▨	■	▨	
All perils except flood, earthquake, war, nuclear accident and others specified in your policy. Check your policy for a complete listing of perils excluded.			▨		■		

▨ — Dwelling and Personal Property
▨ — Dwelling only
■ — Personal Property only

*Included as a peril in traditional forms of the homeowners policy; as an additional coverage in the simplified (HO-76) policies.

are constantly being developed. But here, briefly, are the outlines of today's offerings:

"Basic Policy" (HO-1)

This policy insures the dwelling and other possessions against loss or damage resulting from 11 of the most common perils, on a "named perils" basis (the first 11 perils listed on the chart). Only a small minority of home owners use this form, as it provides less coverage than most people want.

"Broad Form" (HO-2)

Covers you against all 18 perils listed on the chart, and is somewhat more liberal in other respects. Much more widely used than the basic form.

"Special" or "All-Risk" Form (HO-3)

The broadest available coverage for the dwelling and outbuildings, covering loss or damage from any cause except those excluded in the policy (such as war, nuclear accident and others specifically listed). However, furnishings and other personal belongings are covered only for the 18 perils listed on the chart. By last count, this was the most popular of all the homeowners packages, appealing to those who want the widest possible coverage on the home and other buildings, without paying extra for the same extensive coverage on personal property.

A rather complicated example of how this can work is that of the amateur art collector who bought an HO-3 policy, with an added all-risk fine arts floater. A valuable painting that hung over his fireplace fell, breaking three antique clocks (which had not been listed on the floater) and an antique vase (listed), and causing damage to the painting's frame, the plaster wall, the fireplace mantle, the marble hearth and the hardwood floor. A claim was paid against the all-risk floater for the picture frame, and against the all-risk portion of the dwelling coverage for the wall, wallpaper, mantle, hearth and hardwood floor. The vase was covered, but the clocks were not covered, due to the (HO-3) exclusion limiting the "falling objects" peril to objects falling from outside the dwelling.

Renter's Policy (HO-4)

Covers your personal property including furnishings, if you rent a house, apartment or some other residential unit, against the 18 perils listed on the chart. Also covers alterations or improvements you make to the building at your own expense, up to 10 percent of the amount of your personal property limit. Insurance on the building itself is the responsibility of the landlord.

"Comprehensive" Form (HO-5)

Covers both home and personal property on an "all-risk" basis against all but specifically excluded risks. The aristocrat of homeowners packages, and the most expensive.

Condominium Policy (HO-6)

Provides essentially the same coverages as the renter's policy, although with condominium policies "all-risk" coverage is available as an option. Building additions, alterations and decorations you pay for are covered up to $1,000. Like a landlord, the condominium association usually takes care of insurance on the buildings and other structures. Since as a condominium owner you may become subject to assessments for any uninsured property or liability losses to the building, you can cover this risk by an endorsement to your policy.

"Older Home" Insurance (HO-8)

Allows owners of homes whose replacement cost would be higher than their market value to insure them at the lower (market value) amount. In effect, the policy contemplates returning the home to serviceable condition if damaged, but not necessarily with materials of the same kind or quality as the original. Applies to two kinds of dwellings: those whose appointments, such as oak paneling, plaster walls, hand-carved circular staircases and the like, would be exorbitantly expensive to replace today; and those homes whose market value is depressed below the cost to rebuild because of location or neighborhood. A side benefit of the policy is that it removes the possible incentive for arson that exists when a building is over-insured, and thus makes the building a more acceptable risk to an insurance company—helping to solve an insurance availability problem in some declining or "arson-prone" areas. Covers the same 11 perils named in the basic policy (HO-1).

"All-Risk" vs. "Named Perils"

A reason for the popularity of the "all-risk" policy, compared with other homeowners forms, may be found in some actual cases where coverage was, or would have been, provided under the "all-risk" policy (HO-5), but not provided under the "basic" (HO-1) or "broad form" (HO-2) policies.

- A man was walking on floor joists in his attic, when he missed his footing and fell through the ceiling. "All-risk" covered the damage.
- A man laid a fresh cement driveway, which his wife promptly drove over. Covered by "all-risk."
- A visitor went berserk and started throwing furniture and anything else within reach. Damage to the house

was covered, but damage to personal belongings was not covered, as loss did not appear to come under "vandalism and malicious mischief" or any other named peril applying to personal property.
- A Civil War cannonball was dropped into a toilet, which it completely destroyed. Covered under "all-risk," but would not be covered under basic or broad forms because of exclusion on inside falling objects.
- Damage caused by a raccoon which got under a roof and set up housekeeping above a suspended ceiling was covered by "all-risk" in spite of an exclusion for vermin, as a court ruled that raccoons are not vermin.
- Malfunction of a thermostat while a family was on vacation caused the furnace to remain on constantly, resulting in total loss to furnishings, wallpaper, carpeting, books and art works. The loss was covered by "all-risk" (HO-5) but would not have been covered as a fire loss under "named perils" policies as damage was caused by heat without fire.
- Expensive out-of-season clothes stored in cartons were inadvertently given to a charity clothing drive. Covered under "all-risk" (HO-5) but not under "named perils."
- During a party, a drink was spilled into the host's grand piano causing considerable damage, discovered later and necessitated costly repairs. Spilled drinks are not a "named peril," but "all-risk" (HO-5) covered the loss.
- Neighbor children playing in a home owner's finished basement found a hatchet and proceeded to chop up the woodwork. There was no coverage under "named perils" (the owner had (HO-2), but there would have been under "all-risk" (HO-3).

Losses which have been covered by "all-risk" but not by "named peril" policies include property bounced out of an improperly closed car trunk; property lost because of hijacking or mutiny on the high seas; damage to articles in transit, undiscovered until after delivery; many unclassified water losses; and other losses from causes neither named nor specifically excluded.

INDEX

A

Accidents
  personal injury liability, 38, 40, 45, 71–72
  protection after car accident, 157
  under no-fault coverage, 117–124
Adjusters, 69
Age, as auto insurance factor, 130–134
Agents, insurance, *see* Professional advice
Aircraft, 17, 43
Alabama, beach and windstorm plan in, 60
All-risk coverage, personal articles floater, 48–49
American Society of Appraisers, 51
American Stock Exchange, 219–220
Annuities, 169–170
Antiques
  appraisals of, 35
  as investments, 233
  burglary prevention of, 88
  coverage for, 48–50, 58
Appraisals, 19–26, 34, 48, 51, 290
Arbitrators, 69–70, 101
Arbitrary cancellation of insurance, 276
Artworks
  As investments, 233
  burglary prevention of, 88
  coverage for, 48–50, 58, 72
  See also, Inland marine insurance
Auto accidents, 152–158
  aid for victims, 157
  hit-and-run, 157
  put your insurance to work, 155–157
  steps after an accident occurs, 153–155
Auto insurance, 94–151;
  consumer tips, 143–144, 151, 158
  and age, 130–134
  and car types, 139
  and car use, 111–113, 133–139
  bodily injury liability, 98, 115
  borrowed vehicles, 112
  chargeable accidents, 137–138
  claims procedures, 155–157 collision, 101–103, 115, 119

"company cars," 112
comprehensive coverage, 103–104, 111, 115, 119
discounts offered, 141, *142*, 143
drinking and driving, 131, 136
driver classifications 129–140
driver education courses, 135
exclusions, 110–114
gross negligence, 114
how your premium is figured, 127–128
law requirements, 107
medical payments coverage, 99–100, 114
miscellaneous coverages, 104–105
miscellaneous vehicles, 105–107, 116
no-fault (personal injury liability), 117–124
  states with, 120–121
out-of-country requirements, 107–109
points system, 136–137
premium costs, 125–144
property damage liability, 98–99
rating territories, 128–129 rental reimbursement, 105
rent or lease, 113
safety incentives, 139–140
safety records, 135
sex and marital status and rate changes, 134
shared market insurance, 146–151
teenage premiums, 130–133
total replacement, 110
towing and labor, 105
underinsured motorists coverage, 101–115
uninsured motorists coverage, 100–101, 115

B

Bailees customer insurance, 204
Bankers, 141, 272–273
Barron's, 239, 251
Boats, 43, 38
Bests Insurance Rports, 271–272
Bodily injury liability, 98, 115
Bonds, 226–230
  corporate bonds, 229
  convertible bonds, 230
  discount bonds, 229

Fannie Maes, 226
Federal Home Loan Bank Board, 226
Ginnie Maes, 226
Municipal bonds, 226-117
Tennessee Valley Authority (TVA), 226
Treasury bills (T-bills), 226
Treasury bonds, 226
Burglar alarms, 10, 86
   discount for, 32, 35
Burglary, prevention of 86-92,
   consumer tips, 92
   See also, Theft coverage

C

California, domestic
   employee coverage, 43
Canada, out-of-country
   auto insurance, 107-109,
   consumer tips, 116
Census Bureau, 23
Certificates of Deposit, 217, 236
Certified Life Underwriter, 270-271
Certified Property/Casualty
   Underwriter, 270-271
Chicago Board of Options
   Exchange, 232
Child care insurance, 198
Civil commotion coverage, 22, 58
Claims, 67-75, 281-282
consumer tips on, 74-75
   filing an auto insurance claim, 155-157
   filing a home insurance claim, 68-72,
   consumer tips on, 75
   filing a medical claim, 72-73
   filing a property claim, 69-71, 73-74
   settlements costs, 71-75
Claims representatives, 204
Coins, as investments, 231
   protection of, 48, 58
Collectibles, as investments, 233
Collison insurance, 101-103-115,119
Commodities, 232-233
Common stocks, 219
Comprehensive coverage, 103-104, 111, 115, 119
Consumer awareness, 266-282
Consumer tips, 11-12, 24-25, 35,
   44-45, 51, 60-61, 66, 74-75, 92, 114-116,
   124, 143-144, 151, 158, 174-175, 190-191,
   200-201, 209-210, 214-215, 235-238,
   251-253, 264-265
Consumer Price Index, 127, 176
Convertible bonds, 230
Co-ops *see* Real estate
Corporate pension plans, 258-260
Cost of living table, *256*
Credit cards, loss of, 20,68
Crop-hail insurance, 65-66
Crime
   FAIR Plans, 57-60, 61
   government insurance for, 57-61,
   consumer tips, 61
   mugging insurance, 198

D

Dead bolt locks, discounts for, 32, 35
Debris removal
   home insurance, 21
Deductibles
   comprehensive auto insurance, 104
   homeowners policies, 15, 25, 35
   saving money with, 274
   medical policies, 180, 185-186
Dental insurance, 183
Diamonds, 231
Disability benefits, 121
   See also, Health insurance
Disaster
   prevention of loss by, 10-12
   consumer tips, 11
Discount bonds, 229
Discounts
   auto insurance, 141, *142*, 143
   home insurance, 32-33
   safe deposit box storage for, 51
District of Columbia, no fault insurance, 119
Dollar cost averaging, 247-249

E

Earthquakes, insurance against losses
   resulting from, 22, 55-57
   consumer tips on, 61
Economic Recovery Tax Act of 1981
   (ERTA) 221;
   as amended by Tax Equity and
      Fiscal Responsibility Act of 1982, 261

Index 299

Employee savings plans, 260–261
Employment Retirement Income Security Act (ERISA), 258–260
Endowment policies, 165
Evacuation, forced, 19–20
Errors and omissions insurance, 203, 209
Excess and surplus insurance, 194–196

F

Fair Access to Insurance Requirements Plans (FAIR Plans), 57–61,
Family income policies, 165
Family plan policies, 165–166
Family size, changing needs and insurance costs, 278
Farm insurance, 62–66, consumer tips, 66
  all-risk, 65, 138
  basic coverage, 63
  broad coverage, 65
  crop-hail coverage, 65–66
  farm buildings, 63
  farm equipment and livestock, 63–64, 65
  farm personal property, 63, 64
  liability, 62, 64, 66
  medical payments provision, 64
  optional coverages, 64–65
  tenant's broad coverage, 63
Federal Bureau of Investigation, 86
Federal Crime Insurance Program, 59
Federal Crop Insurance Corporation, 65
Federal Deposit Insurance Corporation, 218
Federal Emergency Management Agency (FEMA), 55
Federal Employee's Compensation Act, 208
Federal Employer's Liability Act, 208
Federal Home Loan Bank Board, 226
Federal Insurance Administration, 55
Federal National Mortgage Association (Fannie Mae), 226
  See also, Bonds
Financial security, 160–174
  consumer tips, 174–15
Financial World, 239

Fire 76–82, consumer tips, 92
  coverage under liability, 39
  family escape plan, 79–80
  hazards in the home, 77
  insurance agaist losses caused by, 39, 45
  precautions against, 10, 11
  fireplaces and heaters 81–823
  woodstoves, 80–81
  See also, smoke detectors
Fire department, service charges, 21
Floaters,
  personal articles, 196
Flood insurance, 21–22, 25, 55–57,
  consumer tips, 61
Flood plan management program, 55
Florida, and no-fault insurance, 120
  beach and windstorm plan in, 60
Forbes, 251
Forgery, 20
Fund transfer cards, 21

G

Gains, reinvesting, 249–250
  capital gains, 250
  compounding, 250
  dividend income gains, 250
Georgia, and no-fault insurance, 120
Gold
  as investment, 231
  coverage for, 58
Government National Mortgage Association (Ginnie Mae), 226
  See also, Bonds
Group health insurance, 177–178
Group life insurance, 166
Growth stocks, 221

H

Hawaii, and no-fault insurance, 120
Health Insurance, 176–191, consumer tips, 190–191
  basic coverage, 179–180
  choosing a plan, 183–814
  Consumer Price Index figures on, 176

costs, 187–189
dental insurance, 183
disability income insurance, 181–183
group coverage, 177–178
health maintenance organizations, (HMOs), 178–179
health service contractors, 179
major medical plans, 180–181
specific disease policies, 183
See also, Medicaid and Medicare
Hit-and-run, 157
See also, Auto accidents
Home insurance, 15–16, 20
consumer tips, 24–25, 35
additional property coverage, 20–21
all-risk coverage, 23
amount of coverage, standard, 30
appraisals for, 26–29
business, home-based, 15, 25
comparable costs chart, 31
construction cost coverage, 23, 29
damage to others' property, 40
discounts on, 31–33
forced evacuation, 19–20
full-replacement cost coverage, 27
hazards not covered, 22–23
home appraisal for, 26–29
inflation, 29–20
secondary residences, 18
total loss coverage, 26–27
See also, Homeowners policies
Homeowners policies
all-risk vs. named perils, 295–296
basic policy (HO-1), 294
broad form (HO-2), 294
comprehensive form (HO-5), 295
condominium policy (HO-6), 296
deductibles, 15 exclusions, 21–23 61–62
liability coverage, 36–41
older home insurance (HO-8), 295
renters policy (HO-4), 294
special or all-risk form (HO-3), 294
Home improvements, 29
Home nursing visits, 186
Household appliances, safety of, 77
See also, Inventory household
Hurricanes, 60–61
FAIR Plan coverage, 60
safety precautions for, 84–86

I

Income stocks, 221–222
Individual Retirement Accounts (IRAs), 261–264
Inflation, 49
effects on auto insurance, 127
guards against, 29
Inflation guard endorsement, 30
Injury, see Accident
Inland marine insurance, 196
Insurance, unusual types, 199–200
Insurance agents, 14, 68,
See also, Professional advice
Independent Insurance Agents of America, 270
Insurance companies
changing companies, 278
ratings of companies, 271–272
Insurance policies, 160–175,
appraisal of loss, arbitrary cancellation, 276, 289
company's options, 290
concealment or fraud in, 289
discounts, 275
filing claims, 281–282
how they're put together, 288–290
how to read them, 284–290
insurance insolvency, 276
reassessment of, 276–277
renewal and updating, 277–279
savings features, 161–162
subrogation, 290
waivers, 289
Insurance portfolio, consumer tips, 200–201
Insured money market accounts, 218
Internal Revenue Service, 261
International Association of Chiefs of Police, 91
Inventory
of household items, 33–35, 70, 279–280
of lost, damaged, or stolen articles, 69, 74
Investment clubs, 225
Investment managers, 239–241
Investment trusts, 227–228
Investment, 216–238
Investors Insurance, 197

Index   301

J

Jewelry
  coverage of, 17, 25, 48, 58
  safety of, 35
  See also, Inland marine insurance
Judicare, see Legal expense insurance

K

Kaiser-Permanent, 179
Kansas, and no-fault insurance, 120
Kentucky, and no-fault insurance, 120
Keogh Plan, 242, 263–264

L

Lawsuits
  filed against you, 72
  restrictions of, 122
Legal expense insurance, 186
Liability, sample situations, 71–72
Liability insurance, 36–45, 202, 286–287
  consumer tips, 44–45, 114–116
  auto insurance, 111–124
  bailees customer insurance, 204
  business-related, 41
  called limits, 288
  care for children, 288
  damage to others' property, 40
  exclusions, 41–42, 45
  filing a claim, 204–205
  limits on, 38–39
  malpractice, 41
  medical payments, 40
  personal liability, 37–39, 286–287
  professional liability, 203–204
  rental property, 42–43, 45
  special purpose coverage, 288
  umbrella, 39, 288
  consumer tip, 114
  vehicles and boats, 43
  workers' compensation in the home, 43
  working minors, 42
  See also, Professional liability insurance
Life expectancy table, 256
Life insurance, 106–175
  endowment policies, 165
  family income policies, 165
  family plan policies, 166
  group life insurance, 166

  mortgage cancellation insurance, 166
  policy dividends, 169–171
  savings features, 161–162
  term insurance, 162–164
  universal life insurance, 167
  variable life insurance, 166
  whole life insurance, 164–165
Life insurance riders, 167–169
  cost of living rider, 168
  disability income rider, 169
  double indemnity, 167–168
  guaranteed insurability option, 169
  waiver or premium, 168
Liquid Investment Accounts, 218
Lloyds' of London, 183
Longshoremen's and HarborWorkers' Act, 708
Loss, psychological effects, 9–10;
  controlling your losses,
  consumer tips, 11–12
Loss of use, home insurance 19–20
Louisiana, beach and windstorm plan, 60

M

Malpractice insurance, 41, 203, 209
Marital status, change in, 278
Massachusetts, and no-fault insurance, 120
Medical payments
  auto coverage 99–100, 114
  consumer tips, 114–116
  policyholder liability, 40
  filing a claim for, 72–73
Medical plans, see Health Insurance
Medicaid, 121, 184–187
Medicare, 121, 184–187, 259
  custodial and home nursing visits, 186
  Medigap, 186
Merchant Marine Act, 208
Mexico,
  see Auto insurance, out-of-country coverage, 107–109, 116
Michigan, and no-fault insurance, 119
Minnesota, and no-fault insurance, 120
Mississippi, beach and windstorm plans in, 60
Mobile home owners' insurance, 23–24
Money Management Accounts, 218

Money Market Deposit Accounts, 217–218
Money Market Funds, 245–247
Moody's Investment Service, 224, 229
Mortgage cancellation insurance, 166
Mortgage extra expense insurance, 19
Mortgage interest policy, 298
Motor homes, *see* Auto Insurance, miscellaneous vehicles
Motorcycles,
   insurance of, 105, 111, 116, 120
Mount St. Helen's, 209
Moving and storage, insurance for, 18–19
Mudslide coverage, 21–22, 25
Mugging insurance, 198
Municipal bond insurance, 197
Municipal bonds, 226–227
   long-term andf short-term issues, 227
Mutual funds, 239–253
   consumer tips, 251–253
   advantages of, 242–243, 249–253
   balanced fund, 242
   bond fund, 241
   capital gain fund, 242
   cash reserves, 242
   "closed-end" and "open-end," 243–245
   dollar cost averaging, 247–249
   government reserves, 242
   growth fund, 241
   individual retirement accounts, 242
   Keogh Plans, 242
   load and no-load charge, 243
   preferred fund, 242
   stock fund, 241
   tax-exempt bond fund, 242
   tax-free money fund, 242
   total return fund, 242

N

National Fire Protection Association, 81
National Safety Council, 130
National Weather Service, 81
Negligence, contributory and comparative, 287–288

Neighborhood watch program, 91
New Hampshire
   mandatory comprehensive personal liability insurance, 43
New Jersey
   mandatory comprehensive personal liability insurance, 43
New York, and no-fault insurance, 120
   and theft insurance, 17
New York Stock Exchange, 219–220, 232, 245
No-fault auto insurance, 117–123,
   consumer tips, 124
North Carolina, beach and windstorm plan, 60
Now accounts, 217
Nuclear accidents,
   insurance against losses resulting from, 23, 25
   Three Mile Island, 208–209

O

Officers and directors, insurance for, 198–199
Operation Indentification, 91
Options and commodities, 232–233
Over-the Counter (OTC) Market, 219–220
Overspending, prevention of, 26–35

P

Package policies, advantages of, 275, 285
Pension Benefit Guaranty Corporation, 260
Personal articles floaters, 48–51
   rates, 50
Personal identification program, 33
Personal injury, 36–38, 40, 72
Personal possessions, 46–51
   amounts of coverage for, 47
   floaters, 48–50
   theft of, 46–47, 67–71

Index 303

Personal property coverage, 16–19
  at a secondary residence, 18
  business-related, 18
  exclusions, 17
  filing a claim for, 67–71
  renting to others, 18
  repairs of, 21
Pets,
  as burglary deterrents, 87–88
  as exlcusions on homeowners policies, 17, 25
  health insurance for, 198
  liability for, 37
Points, see Auto insurance, points system
Policies, see Insurance policies
Policy dividends, 170–171
Policy provisions, consumer tips, 51
Pork bellies, 232
Portfolio, protection of, 223–225
Precious metals, as investments, 231
Preferred stocks, 223
Price-Anderson Act, 208
Professional advice, 269–274
  agents, exlcusive and independent, 269–270, 281
  arbitrators, 69–70, 101
  bankers, 141, 272–273
  brokers, 270
  claims representatives, 204
  direct writer representative, 270
  investment counselors, 224
  stockbrokers, 225, 272
Professional Insurance Agents, 270
Professional liability insurance, 203–204
Property damage liability,
consumer tips on, 115
  auto insurance, 98–99
  filing a claim, 69, 71–74
  settlement cost differences, 71, 75
  time limits on, 73–74
  verifying your claim, 69–70
Property Insurance Plans Service Office, 59
Property removal, homeowners insurance, 24

R

Ranch owners, see Farm Insurance
Real estate, investing in, 233–235 as tax advantage, 234
Recession, how to deal with, 212–215
  consumer tips, 214–215
  guarding job skills, 212–213
  inflation guards, 214
  joblessness, 214
  keeping debts down, 213–214
Renters policy (HO-4), 294
Renting
  furnishings, 18
  insurance coverage for, 14–15, 17, 63
Retirement, 254–265;
  consumer tips, 264–265
  alternative plans, 264
  and auto insurance discounts, 131
  and health insurance, 187
  and home insurance discounts, 33, 35
  corporate pension plans, 258–260
  Employment Retirement Income Security Act (ERISA), 258–260
  employee savings plans, 260–261
  IRAs, 261–264
  life expectancy chart, 256
  supplementary retirement plans, 261
Robbery, see Theft coverage

S

Safe deposit box, insurance discount for, 51,
  importance of, 11, 12, 35
Savings deposits, 217
Savings plans, 216–328, 218,
  consumer tips 235–238
Securities and Exchange Commission, 246
Security holdings chart, 227
Settlement costs, 74–75, 204
Shared market insurance, 146–151
  consumer tips, 151
  auto insurance plans, 148
  Joint Underwriting Association, 148
  reinsurance facilities, 148
  state fund, 148–149
Silver, as investment, 231

Smoke detectors, 10, 12, 78–79
  discounts for, 32, 35
Snowmobiles, 106, 116
Social Security, 121, 257–258
  applying for benefits, 258
  work credit benefits table, 257
Social Security Administration, 258
Solar energy systems, discounts for, 33, 35
South Carolina, beach and windstorm plans, 60
Sprinkler systems, discounts for, 32, 35
Stamp Collections, as investments, 233
  protection of, 48, 58
  See also, Inland Marine Insurance
Standard and Poor's Outlook, 224, 228, 232
Statute of limitations, 73–74
Stocks and bonds, as investments, 216–253
Stockbrokers, 225, 272
Subrogation, 290
Surgical survival insurance, 197

T

Tax exempt/taxable yield equivalents, 228
Teenage drivers, 130–131
Tennessee Valley Authority (TVA), 226
Term insurance, 162–164
Texas, and beach and windstorm plans, 60; home insurance discounts, 32–33; tornados, 82
Texas Crime Prevention Institute, 33
Theft coverage, 17, 46–51
  burglaries, 46, 57
  filing a claim, 68–71
Three Mile Island, 208–209
Title insurance, 52–54, 57
  consumer tips on, 61
Tornados,
  insurance against damage caused by
  safety precautions during and after, 82–84
Total return stocks, 222–223
Travel insurance, 196–197
Treasury bills (T-bills), 226
Treasury bonds, 226

Treehouse coverage, 24
Trees, shrubs and plants, and home insurance, 21

U

Umbrella coverage, 39, 114
Underinsured motorist coverage, 101–115
Underwriter's Laboratories, 78, 81
Unemployment Insurance for executives, 199,
United States Bureau of Labor Statistics, 127
United States Department of Agriculture 65
United States Dept. of Labor, 176
Universal life, 167
Uninsured motorists coverage, 100–101, 115

V

Value Line Investment Service, 224
Vandalism,
  and farm insurance, 63
Variable life insurance, 166
Victims, aide for, 157, 202–210
  consumer tips 209–210
Violent crime victims insurance, 198
Volcanic eruptions, insurance against losses caused by, Mount St. Helens, 209

W

Wall Street Journal, The 250
Water damage exclusion, home insurance, 22
Whole life insurance, 164–165
Windstorms, consumer tips, 92
  beach and windstorm coverage, 60
  safety precautions in the event of, 82–83
  See also, Tornados and Hurricanes
Workers' compensation, 202, 203, 205–208
  and auto accidents, 121
  benefits, 207
  home insurance coverage, 43
  options for employers, exceptions to payments, 207–208
  special insurance pools, 208
Working minors, 42
Wood stoves, safety factors, 80–81